HIERONYMI CARDANI, PRÆSTANTISSIMI MATHEMATICI, PHILOSOPHI, AC MEDICI,

ARTIS MAGNÆ,

SIVE DE REGVLIS ALGEBRAICIS,

Lib. unus. Qui & totius operis de Arithmetica, quod

OPVS PERFECTVM

inſcripſit, eſt in ordine Decimus.

HAbes in hoc libro, ſtudioſe Lector, Regulas Algebraicas (Itali, de la Coſſa uocant) nouis adinuentionibus, ac demonſtrationibus ab Authore ita locupletatas, ut pro pauculis antea uulgò tritis, iam ſeptuaginta euaſerint. Neq; ſolum, ubi unus numerus alteri, aut duo uni, uerum etiam, ubi duo duobus, aut tres uni ęquales fuerint, nodum explicant. Hunc aũt librum ideo ſeorſim edere placuit, ut hoc abſtruſiſsimo, & planè inexhauſto totius Arithmeticæ theſauro in lucem eruto, & quaſi in theatro quodam omnibus ad ſpectandum expoſito, Lectores incitarẽtur, ut reliquos Operis Perfecti libros, qui per Tomos edentur, tanto auidius amplectantur, ac minore faſtidio perdiſcant.

ARS MAGNA
or
The Rules of Algebra

GIROLAMO CARDANO

Translated and Edited by
T. RICHARD WITMER
With a Foreword by Oystein Ore

DOVER PUBLICATIONS, INC.
New York

Copyright

Published in Canada by General Publishing Company, Ltd., 30 Lesmill Road, Don Mills, Toronto, Ontario.
Published in the United Kingdom by Constable and Company, Ltd., 3 The Lanchesters, 162-164 Fulham Palace Road, London W6 9ER.

Bibliographical Note

This Dover edition, first published in 1993, is an unabridged republication of the edition published by The MIT Press in 1968 under the title *The Great Art or The Rules of Algebra.* It was translated and edited from the 1545 edition of *Artis magnae, sive de regulis algebraicis. Lib. unus. Qui & totius operis de arithmetica, quod Opus Perfectum inscripsit, est in ordine decimus* with additions from the 1570 and 1663 editions. This Dover edition is published by special arrangement with The MIT Press, 55 Hayward Street, Cambridge, Massachusetts 02142.

Library of Congress Cataloging-in-Publication Data

Cardano, Girolamo, 1501–1576.
[Ars magna. Liber 1. English]
Ars magna, or, The rules of algebra / Girolamo Cardano ; translated and edited by T. Richard Witmer ; with a foreword by Oystein Ore.
p. cm.
Originally published: The great art, or The rules of algebra. Cambridge, Mass. : M.I.T. Press, 1968, which was translated and edited from the 1545 ed. of Ars magna, liber unus, with additions from the 1570 and 1663 eds.
Includes bibliographical references and index.
ISBN 0-486-67811-3 (pbk.)
1. Algebra—Early works to 1800. I. Witmer, T. Richard, 1909– II. Title. III. Title: Ars magna. IV. Title: Rules of algebra.
QA154.8.C3713 1993
512.9′4—dc20 93-34446
CIP

Manufactured in the United States of America
Dover Publications, Inc., 31 East 2nd Street, Mineola, N.Y. 11501

Table of Contents

Foreword

It has often been pointed out that three of the greatest masterpieces of science created during the Rinascimento appeared in print almost simultaneously: Copernicus, *De Revolutionibus Orbium Coelestium* (1543); Vesalius, *De Fabrica Humani Corporis* (1543), and finally Girolamo Cardano, *Artis Magnae Sive de Regulis Algebraicis* (1545). But while the two first works have been readily available in magnificent editions and excellent translations, Cardano's *Ars Magna* has remained relatively obscure, its material confined to early and now rare Latin editions. The *Ars Magna* has always been highly praised as a milestone in the history of mathematics, yet it is true that the number of modern scholars who can claim to have examined it in detail is extremely small. Thus the present translation, making it available to a much wider circle of readers, is a most significant addition to the literature of the history of science.

The creation of a modern version of an ancient scientific text poses a number of problems—the suitable choice of new or old technical terms, the rendering and interpretation of obscure or confused passages, and also, peculiar to mathematics, the decision to which extent one should introduce present-day symbolism. These matters of judgment seem to have been happily resolved in this translation. Cardano's work, cumbersome as it is in its original form, complicated by the rudimentary mathematical language of the period, now emerges as a book which can be studied by any college undergraduate. The systematic use of ordinary school algebra in Cardano's reasonings contributes greatly to this simplification.

Nowadays, the *Ars Magna* would be characterized as a text on algebraic equations. To Cardano's contemporaries it was a breakthrough in the field of mathematics, exhibiting publicly for the first time the principles for solving both cubic and biquadratic equations, giving the roots by expressions formed by radicals, in a manner similar to the method which had been known for equations of the second degree since the Greeks, or even the Babylonians. Cardano actually does not claim either of the two innovations entirely as his own; he rather considers the special third degree equation first solved by Scipione del Ferro, and the fourth degree equation solved by Lodovico Ferrari, Cardano's secretary, as toeholds which enable him to create his own general theory embracing all possible cases.

These many cases, produced largely by the necessity for separating the arguments for positive and negative numbers and by the lack of an efficient algebraic notation, lead to the elaborate lists of equation types. Cardano studies what he considered to be properties of general equations, for instance, relations between roots and coefficients, rules for the signs or the locations of the roots. He barely touches upon the numerical solution of equations, but here he brings little new. One notable aspect of Cardano's discussion is the clear realization of the existence of imaginary or complex solutions. They appear as necessary consequences of the formulas, and he does not avoid them or brush them aside as unimportant, as often done by earlier writers. On the contrary, Cardano constructs examples for the express purpose of dealing with problems with imaginary roots (Chapter 37).

In dealing with problems as complicated as the solution of higher degree equations, it is evident that Cardano is straining to the utmost the capabilities of the algebraic system available to him. Shortly afterwards his successors set to work to create a more general and more efficient algebraic language. This is noticeable already in Bombelli's *Algebra* (1572); in 1591 appeared Viela's work, *In Artem Analyticam Isagoge*, which brought mathematical terminology to a stage approaching the modern one.

Cardano complicates his notational difficulties further by basing most of his proofs upon geometric arguments, thus emulating the reasoning of Euclid as he repeatedly emphasises. To us there is no necessity for the use of such methods; on the contrary, they appear strongly incongruent in this connection. But this was a period of the highest veneration of the methods of Greek mathematics, and Euclid's *Elements* represented the pinnacle of logical stringency. Euclid's geometric solution of second degree equations consisted in constructing squares, and so the proper approach to cubic equations would be through the construction of cubes. But from then on the guidance of the geometric intuition disappeared, and it is curious to note that Cardano seems to feel that there is a connection between the fact that space as he knew it was three dimensional, and the difficulties in solving higher degree equations: "Nature does not permit it" (Chapter 1).

The *Ars Magna*, in spite of its novelty, is influenced to some extent by the centuries of Italian algebra from Leonardo Pisano's *Liber Abaci* (1202) to the *Summa* (1494) of Fra Luca Paccinolo. As in the lectures of the Abacista, the problems are often clothed in practical garb to make them more attractive to the readers. For equations of higher degree this is no easy task, so in most cases Cardano falls back upon straight numerical examples. But a number of the examples are of the types reminiscent of mercantile Italian arithmetics of late medieval times. Among them let us mention the distribution of monies among soldiers, profit on repeated business trips, and the ever present questions concerning partnerships (*regola della compagnia*, Chapter 5). All these problems have a highly artificial aspect. The only ones in which the cubic equations appear in nearly natural fashion are the extension of the Delian problem concerning the doubling of the cube (Chapter 17), geometric problems concerning triangles (Chapters 32, 38) and questions concerning compound interest (Chapters 18, 20). Chapter 37 is one of the few where Cardano deals freely with negative numbers. One of the examples is quite illustrative, and this situation was probably not rare. The dowry of the bride is compared to the estate of the husband, and it turns out in the end that he is so heavily in debt that the dowry barely bails him out.

The *Ars Magna*, besides being an outstanding mathematical achievement, is also famous as the direct cause of one of the most violent feuds in the history of science. Cardano acknowledges abundantly, in fact three times (in Chapters 1, 6, and 11) that he had originally received the solution to the special cubic equation

$$x^3 + ax = b$$

from his friend Niccolò Tartaglia of Brescia, at the time *abacistà* in Venice. Cardano's account in Chapter 11 runs as follows: "Scipio Ferro of Bologna well-nigh thirty years ago discovered this rule and handed it on to Antonio Maria Fior of Venice, whose contest with Niccolò Tartaglia of Brescia gave Niccolò occasion to discover it. He [Tartaglia] gave it to me in response to my entreaties, though withholding the demonstration." This places the original discovery of the solution by Scipione del Ferro, professor of mathematics in Bologna, at approximately the year 1515. Scipione died in 1526, and at that time his manuscripts passed to his son-in-law and successor Annibale della Nave. These facts are fully confirmed by the rediscovery by E. Bartolotti of del Ferro's original papers in the library of the university of Bologna.

At his death the solution was known by certain of del Ferro's pupils, certainly by Annibale della Nave and Antonio Maria Fiore. The latter returned to his native town of Venice, probably to make a living as a teacher of mathematics. To make his name known, he in 1535 challenged Tartaglia to a public problem-solving contest. Fiore's problems all concerned del Ferro's solution of the equation: The cosa and the cube equal to a number. Tartaglia describes his agony before the contest, but in the nick of time, the night before the contest, he succeeded in finding the method, and Fiore suffered a humiliating defeat.

Word of the contest reached Cardano in Milan who at the time was preparing the manuscript for his *Practica Arithmeticae Generalis*, which appeared in 1539. Cardano had obviously hoped that Tartaglia would divulge the secret and give him permission to include the method for the cosa and the cube in his book. This Tartaglia flatly refused with the statement that in due time he would write a book on the subject.

So far the facts seem clear, but for the subsequent events we are dependent almost exclusively upon Tartaglia's printed accounts, which by no stretch of imagination can be regarded as objective. It appears that not long afterwards Tartaglia had at least a partial change of heart. He accepted an invitation from Cardano to visit him in Milan, possibly in the hope that through Cardano's influence he could secure a position as advisor to Alfonso d'Avalos, the Spanish viceroy and commander in chief in Milan. Tartaglia had written a book on artillery and constructed instruments for measuring distances in the field. At a meeting in Cardano's house the secret was divulged and, according to Tartaglia, Cardano swore a most solemn oath, by the Sacred Gospels and his word as a gentleman, never to publish the method, and he pledged by his Christian faith to put it down in cipher, so that it would be unintelligible to anyone after his death. At this point the word of one man stands against that of another. Ferrari, then a youth of eighteen years and a servant in the Cardano household, related later that he was present at the meeting, and there had been no question of secrecy involved.

After the publication of the *Ars Magna*, Tartaglia's rage knew no bounds. Already the following year he published another book *Quesiti et Inventioni Diverse* in which he exposed Cardano's perfidy, reproducing the correspondence between them, and giving a word-for-word account of their conversations. On February 10, 1574, Ferrari responded with a printed *cartello*, a challenge to Tartaglia to meet him in a dispute on almost any scientific topic, for a prize of up to 200 scudi. "This I have proposed to make known, that you have written things which falsely and unworthily slander the above-mentioned Signor Gerolamo (Cardano), compared to whom you are hardly worth mentioning."

Ferrari's challenge was sent to a large number of scholars and prominent persons all over Italy. Nine days later Tartaglia had his counter statement ready. Altogether six cartels and corresponding countercartels were exchanged between the two during the following year and a half. Tartaglia, as the insulted part in this scholarly duel, claimed the right to choose the weapons, that is, the topics to be discussed, and he explicitly excludes the *Ars Magna*. He repeatedly insists on facing Cardano, then already a famous man, and not Ferrari, the pupil. To malign Cardano's character, possibly he hoped for censure by some of the church dignitaries, Tartaglia repeatedly reminds the readers of his breach of solemn oaths.

Ferrari on the other hand does not play second fiddle to Tartaglia in heaping ridicule and grotesque insults upon the opponent, but he does make certain important points, for instance: "Signor Gerolamo has been able to attribute the result to its first inventor Scipione del Ferro of Bologna, and besides him also to Maria Fiore, who knew it before you, as you confess in your book; nevertheless he has been so courteous that he wanted to believe also that you had found it without having received it from either one of them, or from any of their pupils."

Tartaglia at one stage interrupted the exchange and seemed to wish to withdraw. But at the time he was negotiating for a public lectureship in his native town of Brescia, and the city fathers may have insisted on his participation to prove his qualifications. Anyway, all of a sudden Tartaglia declared his willingness to dispute wherever it may be, even in Ferrari's home town of Milan. The meeting took place in a church in Milan before a great audience on August 10, 1548. We have no account of the proceedings, except a few statements from Tartaglia to the effect that the meeting broke up when the supper hour drew near; the public may also have found the arguments a bit tedious.

From various inferences there does not seem to be much doubt that Ferrari was declared the winner. He received numerous distinguished offers afterwards, among them a request to serve as the tutor for the emperor's son. He died as professor of mathematics in Bologna. Tartaglia did not receive the promised stipend in Brescia and wrote a bitter book also on this subject, describing in detail the injustices meted out to him by the people of the town. Had he been the winner of the dispute in Milan, he would undoubtedly have mentioned this fact.

To conclude, what moral judgment can be passed upon the Cardano–Tartaglia controversy? Tartaglia evidently felt that he had been cheated out of a most precious possession, possibly also of economic value in attracting pupils to his school. On the other hand, Cardano and Ferrari had an excellent counterargument which may even have been unknown to Tartaglia at the time he proffered his charges. In 1543, when the *Ars Magna* neared its completion, Cardano and Ferrari had traveled together to Bologna and from Annibale delle Nave received permission to inspect del Ferro's papers. Here they found the original solution, antedating that of Tartaglia by twenty years, and they maintained that this was the discovery reproduced in the *Ars Magna.*

To a modern scientist the whole incident may appear as an amusing but not too important affair. Perhaps he may consider it a further confirmation of his "Publish-or-Perish" doctrine, and most would feel that Tartaglia had received due credit by being mentioned several times as a codiscoverer. It is, however, entirely inappropriate that the formula for the solution of the simplified cubic equation is still known as Cardano's formula; mathematical writers should make a concerted effort to introduce the name: del Ferro's formula.

The mathematicians in the period following Cardano had high praise for the *Ars Magna*, and most of them do not even refer to Tartaglia, although in a couple of places he is mentioned as an irascible person whose charges carry little weight. But in the subsequent centuries the attitude of the writers on Cardano changed markedly. His reputation in his own days as a universal genius was replaced by a highly critical, even malicious, assessment of his character. His sanity has been questioned, it was pointed out that his numerous works also included astrology and horoscopes, magic and superstitions, as well as many impious arguments on religion which in the end brought him to trial and conviction before the Inquisition. The vitriolic charges by Tartaglia fitted perfectly into this picture. Presently the historians of science have again changed to a much more appreciative view of Cardano, realizing his great role as an innovator, particularly in mathematics and medicine, and also that some of his critics must be taken with many grains of salt.

The second edition of Cardano's *Ars Magna* appeared in 1570, and late in the same year Cardano was arrested by the Inquisition and jailed in Bologna. It seems possible that there might be some vague connection between the two happenings. The charges against Cardano are not known, and many conjectures have been proposed. Certainly it was not because of Tartaglia's charges of breach of his Christian faith, Tartaglia and Ferrari were both dead, and this matter was an everyday transgression of no systematic importance for the Catholic faith. However, this was the time of the Counterreformation, and one of the items specified during the process may well have been Cardano's cordial dedication to Andreas Osiander. Not only had Osiander been one of the aggressive leaders within the German Reformation, but he was also known to have written the preface to Copernicus' *De Revolutionibus Orbium Coelestium*, a work under strong suspicion of being heretical.

Oystein Ore

New Haven, *Connecticut*
July, 1968

Preface

I

The Great Art was the first product of the renaissance in algebra that swept Europe, and particularly the valley of the Po River, in the early sixteenth century. Compared with where we are today in mathematics, the contents of the book are elementary and, in some respects, even crude. Compared with where algebra was fifty or, for that matter, even ten years before the book was published, its importance in opening up a long-stagnant field of learning and inquiry cannot be overestimated.[1]

Although only the name of Girolamo Cardano appears on the titlepage of *The Great Art*, the new materials which it contained were, in a very real sense, the product of four men's work. Two of these—Scipione del Ferro and Niccolò Tartaglia—contributed their basic discovery of a solution for one or perhaps two of the thirteen forms the cubic equation can take.[2] To another, Lodovico Ferrari, a servant, pupil, and collaborator of Cardano's, belongs the credit for discovery of a method that is still in use for solution of the biquadratic equation.[3]

The fourth man, of course, was the author of the book. It was he who developed the proof that the formula or formulae that he received from

[1] For a 50-year comparison, see Luca Paccioli's *Sũma de Arithmetica Geometria Proportioni & Proportionalita*, page 8, footnote 4 *infra*. For a 10-year comparison, see Cardano's own *Practica Arithmeticae*, note 18 *infra*, in which the only cases of the cubic discussed are those which the author can readily manipulate in such fashion as to discover a common denominator for the two sides of the equation and, applying this divisor, can reduce to a quadratic. He points out, for example, that this can easily be done in any case of $x^3 = ax + N$ or $x^3 + N = ax$ if $a = N + 1$, or $2a = N + 8$, or $3a = N + 27$, etc.

[2] See pp. xviii ff. and footnote 11, p. 239 *infra*.

[3] See pp. 237 ff. *infra*.

del Ferro and Tartaglia are correct;[4] found the method for reducing the more complex forms of the cubic—those which include all four possible terms and those three-term cases involving a quadratic term—to one or another of the simple forms and thereby expanded del Ferro's and Tartaglia's discovery to cover all possible forms of the cubic;[5] found and demonstrated the existence of multiple roots for various types of the cubic and biquadratic;[6] explored and set out the relations between the roots of one type of cubic and those of other types directly related to it and between these roots and the given numerical terms of the equations involved;[7] showed that in some instances an equation may have two identical roots;[8] developed consciousness of the importance and inevitability of negative solutions;[9] insisted on the same for complex and irrational numbers;[10] and found himself led to solutions involving the square root of negative quantities, to some extent showing how to manipulate them in spite of what he called their "sophisticated" nature.[11] All this he took the trouble to put down on paper and publish with profit not only to his own generation but to generations that followed. In doing this he enabled the algebra which, as he began his work, was still pretty much earth-bound to sprout

[4] See particularly Chapters XI and XII of *The Great Art.*

[5] See particularly Chapters VII and XIIII–XXIII of *The Great Art.*

[6] See particularly Chapters I and XIII of *The Great Art.* Chapter I, the reader will note, is very different in style and content from most of the rest of the book, consisting as it does almost exclusively of a series of assertions concerning the number of solutions each type of equation can have without offering proof for any of these assertions. To find proof of any kind in any case of the cubic we have to go to Chapter XIIII, where Cardano deals with the subject in connection with $x^3 + N = ax$.

Though the importance that Cardano rightly attached to his discoveries in this field is indicated by the fact that he gave them first place in his book, he left no hint as to what it was that led him to investigate the subject. The del Ferro-Tartaglia formula does not lead one there. From the fact that Cardano's proof is given in the case just mentioned, however, we can perhaps surmise that he was led to it by his knowledge that al-Khowarizmi had discovered that $x^2 + N = ax$, the parallel case in the quadratic, has two roots (see Chapter V). Or we may guess that he came to it when he discovered, in the course of synthesizing an equation for use in his book (say $x^3 + 9 = 12x$, which is deliberately constructed to give 3 as an answer), that his own rule for solving it would yield a totally different result. Beyond this initial discovery, Cardano's exposition in Chapter I seems to rest partly on keen observation and partly on the process he follows for reducing other forms of equations to one or another of the three simple types of the cubic.

[7] See for instance, paragraphs 5, 6, 8, 9, 10 and 11 of Chapter I and *passim* in Chapter IIII.

[8] See pp. 12–13 *infra.*

[9] See Chapter I, *passim.*

[10] See particularly Chapter IIII.

[11] See Chapter XXXVII.

wings and to go where the "realists" of his day and even he himself imagined it could never go.[12]

II

Girolamo Cardano (born 1501 in Pavia, died 1576 in Rome)[13] was a prolific writer on a wide variety of subjects—medicine, astronomy, astrology, philosophy, mathematics, and others. When his extant works were collected and published in 1663, they filled ten large double-column folio volumes which, even so, did not include everything that he had written. Many of his manuscripts had disappeared in the meantime. We know, for instance, from his autobiography[14] and from *De Libris Propriis*[15] that he had written a number of books on mathematical subjects that never saw the light of day. *The Great Art* itself was the tenth in a series of projected volumes in which the author intended to cover the entire field of arithmetic, algebra, and geometry.[16] Though all, or nearly all, of these volumes were actually written, only the ones on proportions and on properties of numbers, and part of the one on integers, have survived.[17] In addition to these, Cardano produced a *Practica Arithmeticae*;[18] a work on what he called the aliza problem;[19] another entitled *Ars Magna Arithmeticae*;[20] a small

[12] See p. 9 *infra*.

[13] For lives of Cardano in English, see his autobiography *The Book of My Life*, tr. Jean Stoner (New York, 1930); James Eckman, *Jerome Cardan* (Baltimore, 1946); Henry Morley, *Jerome Cardan* (London, 1854); Oystein Ore, *Cardano, the Gambling Scholar* (Princeton, 1953); William G. Waters, *Jerome Cardan* (London, 1898).

[14] *The Book of My Life*, note 13 *supra*, pp. 220 ff.

[15] Three versions of this book, one prepared in 1543, another dated 1554, and a third that is undated, are printed in Cardano's *Opera Omnia*, vol. I (Lyons, 1663). His mathematical works are listed on pp. 66, 74 ff. of this volume.

[16] This series is referred to as *The Perfect Work* on the title page of *The Great Art*. In the 1554 version of the *De Libris Propriis*, Cardano tells us that the first book of the series covered integers, the second fractions, the third square and cube roots, the fourth unknowns, the fifth proportions, the sixth properties of numbers, the seventh commercial problems, the eighth principal and interest, the ninth "certain extraordinary matters," the tenth *The Great Art*, the eleventh the measurement of planes, the twelfth the measurement of solids, the thirteenth arithmetic problems, and the fourteenth geometric problems. This list differs somewhat from that given in the 1543 version of *De Libris Propriis*.

[17] The first two of these appear in volume IV of the *Opera Omnia*, the third in volume X. In addition, the *De Proportionibus* was printed in 1570 in the same volume as that year's edition of *The Great Art*.

[18] First published in Milan, 1539; reprinted in *Opera Omnia*, vol. IV.

[19] *De Regula Aliza Libellus*. Like the *De Proportionibus*, this work was printed in 1570 in the same volume as that year's edition of *The Great Art*.

[20] It is not possible to date this interesting little work precisely. Internal evidence,

book on games of chance;[21] a number of commentaries on Euclid which were never published;[22] and quite a few minor compositions, some of which were published and some of which were not.

In terms of permanent significance, *The Great Art* undoubtedly stands at the head of the entire corpus of Cardano's writings, mathematical or otherwise. Contrary to some popular impressions,[23] Cardano did not claim credit for all the new material that is set out in his book. This the book makes quite clear at various places. He was no solitary, lonely genius remote from the intellectual world of his day. As a teacher of mathematics and a writer on the subject, he was interested in expounding all the new ideas he could lay his hands on, not only his own but those of other men as well, and he would probably have considered himself derelict if he had not done so. It was this attitude as opposed to the secretiveness of Tartaglia that led to conflict between the two.

Cardano knew, before he ever started writing *The Great Art*, that there had been a breakthrough in the cubic and that at least two people were in possession of some formula or formulae dealing with it with which he was not familiar.[24] He persuaded one of them, Niccolò Tartaglia, to let him in on the secret—perhaps under promises that he would never reveal it to anyone else, as Tartaglia said;[25] perhaps without any such promises, as Ferrari contended[26]—and he and

---

however, suggests that it was written after *The Great Art*, and its dedication to Filippo Archinto, bishop of Borgo San Sepolcro, indicates that it was completed no later than 1554, the year Bishop Archinto left that post.

[21] *The Book of Games of Chance*, tr. Sydney Henry Gould (New York, 1961). The same is also printed with Oystein Ore's *Cardano, the Gambling Scholar*, note 13 *supra*.

[22] In the 1554 version of *De Libris Propriis*, Cardano tell us that he had completed fifteen books on the *Elements*. In the *Book of My Life*, note 13 *supra*, p. 222, he lists two books entitled *New Geometry*.

[23] For an extreme example, see footnote 8, page 8, *infra*.

[24] See footnote 11, page 239, *infra*.

[25] For Tartaglia's account, see Book IX of his *Quesiti et Inventioni Diverse* (facsimile reproduction of 1554 edition, Brescia, 1959), *passim* but particularly Quaesito XXXIIII, pp. 120–121. Professor Ettore Bortolotti, in his *La Storia della Matematica nella Università di Bologna* (Bologna, 1947), p. 49, expresses strong misgivings about the accuracy of Tartaglia's account, first published in 1546, or seven years after the affair referred to by Tartaglia took place, and questions the authenticity of the purported letters from Cardano to Tartaglia which appear in the latter's book.

[26] For Ferrari's account, see his second letter to Tartaglia (April 1, 1547), particularly pp. 2–3, which is reprinted in facsimile in *I Sei Cartelli . . . di Lodovico Ferrari, coi Sei Contro-cartelli in Risposta di Nicolò Tartaglia*, ed. Enrico Giordani (Milan, 1876). In brief, Ferrari says that he was present in Cardano's house when Tartaglia gave Cardano

Ferrari later discovered the same or a very similar formula in the handwriting of Scipione del Ferro, the man who had first made the breakthrough.[27]

Be the promise or no-promise story as it may, and leaving to one side the probability that Cardano would eventually have found del Ferro's manuscript or that, merely knowing that a solution existed, he would in time have worked it out for himself, there is no question that Tartaglia's revelation of the formula for solving cases of the $x^3 + ax = N$ type furnished Cardano with a very important lead. When Tartaglia answered Ferrari's statement that Cardano had rescued his "little discovery" from oblivion, planted it in a rich garden, nurtured it, expanded on it, and made the name of Tartaglia famous[28] by exclaiming that, but for his contribution, Cardano's garden would have remained a forgotten woodlot,[29] he was exaggerating, but behind the exaggeration was a grain of truth.

It is easy to understand how Tartaglia felt about Cardano's publication of his discovery. Quite apart from its value to him in contests with other mathematicians of the time, he had not worked out all its ramifications. He wanted to and intended to work these out and to publish them under his own name. This, he thought, precluded him from teaching his formula even to his pupils and from disclosing it to his friends—to all except one, that is. But he allowed time to slide by

---

his *inventiunculam*, that no promises were made, and, in effect or by implication, that the gift was in return for Cardano's hospitality. In assessing the value of this statement, it must not be forgotten that Ferrari was a pupil and a strong partisan of Cardano's. In this same letter (p. 5), he speaks of Cardano as "one to whom I owe everything."

[27] *Anno ab hinc quinto, cũ Cardanus Florentiã proficisceretur, egoq; ei comes essem, Bononiae Annibalem de Nave virũ ingeniosum, et humanũ visimus, qui nobis ostendit libellũ manu Scipionis Ferrei soceri sui iã diu cõscriptũ, in quo istud inventũ elegãter et docte explicatũ, tradebatur.* So wrote Ferrari to Tartalgia on page 3 of the letter referred to in the preceding footnote in a context in which he, in effect, accused Tartaglia of having passed on to Cardano, as his own discovery, the discovery of another. Scipione del Ferro's exposition of his discovery, long lost, was returned to light some years ago by Professor Ettore Bortolotti (see his "Manoscritti Matematici Riguardanti la Storia dell'Algebra, Esistenti nelle Biblioteche di Bologna," in *Esercitazioni Matematiche di Catania*, III, 81 [1923] and his *Storia della Matematica nella Università di Bologna*, note 25 *supra*, pp. 43 ff.).

[28] In this same second letter (p. 3), Ferrari said: *Cardanus ergo ex te accepit inventiunculam illam cubi & laterum aequalium numero, quam ut ab interitu, cui vicina erat revocaret, in subtilissimo atque eruditissimo suo volumine, velut languentem & semimortuã arbusculam in amplissimo, feracissimo, & amoenissimo horto inservit, te inventorem celebravit, te exoratum sibi tradidisse commemoravit. Quid vis amplius?*

[29] Tartaglia's response (pp. 5–6 of his second letter) was: *Non vedeti voi che cavando la detta mia pianta del vostro giardino, tal vostro giardino restaria una oscura selva, perche tutte le altre cose sostiantiale derivano da detta mia pianta.*

while he pursued other subjects.[30] Now this other person had done the work he had cut out for himself, published the results, and thereby gained the glory that, Tartaglia thought, should have been his. No matter that he was given credit for the basic formula in his acquaintance's work. This is not the same thing as having a book, an original work, under your own name. It is the book, not the footnote reference, that men will cite. Moreover, though he was given credit for having discovered the principal formula, it had become but one of a baker's dozen of formulae and his acquaintance claimed credit for all the rest of them.

Yet we can also appreciate Cardano's position. For the plain fact is that Tartaglia's giving him the formula for $x^3 + ax = N$ or even the formulae for this and for $x^3 = ax + N$[31] did not automatically provide

[30] *Quesiti et Inventioni*, note 25 *supra*, p. 120r: *Io ve diro, io non fazzo tante il carestioso, per il simplice capitolo, ne per le cose ritrovate per lui, ma per quelle, che per notitia di quello si possono ritrovare, perche eglie una chiava, che ne apre la via à potere investigare infiniti altri capitoli, & se il non fusse che al presente io son occupata nella tradutione di Euclide, in volgare . . . à molti altri capitoli haveria gia trovato regola generale, ma spedito che habbia questa mia fatica di Euclide, gia principiata, ho designato di cõponere un'opera di practica, & insieme con quella una nuova Algebra, nella quale non solamente ho deliberato di publicare ad ogni huomo tutte la dette mie inventioni de capitoli nuovi, ma molti altri, che spero di ritrovare . . . & questa é la causa, che me gli fa negar ad ogniuno. . . .* [*T*] *ale mie invẽtioni le voglio publicare in opere mie, et non in opere de altra persona.*

[31] Tartaglia's account in his *Quesiti et Inventioni*, p. 120v, leaves the impression that he had turned over to Cardano the formula not only for these two cases but for that of $x^3 + N = ax$ as well. If Morley (*supra* note 13, I, 250–251) is correct in translating the last part of the verses in which Tartaglia set out his formulae —

El terzo [$x^3 + N = ax$] *poi de questi nostri conti*
Se solve col secondo [$x^3 = ax + N$] *se ben guardi*
Che per natura son quasi congionti —

as meaning that the solution for $x^3 + N = ax$ is the same as that for $x^3 = ax + N$, Cardano was given something worthless. As to the other two, there is no doubt that Tartaglia furnished Cardano with his formula (but not, Cardano points out, with its proof) for $x^3 + ax = N$. Considerable doubt is cast on the implication that he also handed him the formula for $x^3 = ax + N$ by (1) the fact that Cardano is careful, not only in Chapters I and XI of *The Great Art* but also in Chapter XXVIII of the *Ars Magna Arithmeticae* to express his indebtedness to Tartaglia with respect to the first but not with respect to the second; and (2) Ferrari's specific reference to his witnessing Cardano's acceptance of Tartaglia's "little discovery of the [solution for the] cube and first power equal to a constant" without mentioning the other (see the quotation from Ferrari's letter to Tartaglia, note 28 *supra*) and Tartaglia's constant use of the singular in his reply to this and his failure to claim that Ferrari had understated his (Tartaglia's) contribution in this respect. On the other hand, see Cardano's account in the *De Libris Propriis*, quoted note 11, page 239, where he seems to acknowledge receiving both formulae, and, assuming it is authentic, his letter to Tartaglia dated August 4, 1539 (*Quesiti et Inventioni*, p. 122v), commenting on the solution for $x^3 = ax + N$ when $(a/3)^3 > (N/2)^2$. Whether he received both of them

solutions for all the other forms the cubic equation can take.[32] Men had not yet learned the trick of setting all terms on one side of the equality sign with a zero on the other, and they had not yet learned that missing terms are not missing but merely have a zero coefficient. These were the two conditions that had to be met before there could be, properly speaking, *a* formula for solving *the* cubic.[33] In fact, there was no one cubic at that time. There were thirteen of them, and the formula for each of the thirteen had to be devised and proved separately. This is what Cardano did. It would have been surprising, therefore, if he had not felt entitled to publish the results of his labors, surprising if he had thought that his promise (assuming he did, in fact, make one) foreclosed him forever from publishing not only the formula he was given but everything that could be said to have stemmed from it regardless of his own contribution, and particularly surprising if he had so thought after he discovered that Tartaglia was not the only one who possessed the basic formula and that it had been written down by Scipione del Ferro for the benefit of posterity. Not only would it have been surprising, but the world would have been that much poorer for it since Tartaglia, even though he had had a lengthy headstart on Cardano, never got around to completing the work he had hoped to finish and

---

or only one would not be of great moment in today's state of the art, but it would be anachronistic to say that it was of no importance at the time in which the two men were working. It is interesting, in any event, to notice how much less fruitful Cardano found the del Ferro-Tartaglia type than the other two. Of the ten forms of the cubic involving a quadratic term, he reduces only two to $x^3 + ax = N$, and then only in special circumstances. All the rest reduce to $x^3 = ax + N$ either directly or indirectly through the intermediate step of $x^3 + N = ax$.

[32] Tartaglia continually claimed to have formulae for the three-term cases which involve a quadratic term in place of a linear. See, for example, his *Quesiti et Inventioni*, pp. 101v, 106v, 126v, among others, and his second reply to Ferrari, p. 6, in *I Sei Cartelli: Et tamen el non se vergogna de dire nella detta sua opera, che tutti li altri capituli che in quella si trovano oltre il mio esser tutte sue & vostre inventionui le quale erano state da me invente, & ritrovate gia .5. anni avanti che gli insegnasse a lui tal mia particolarita, come che e noto a molti qua in Venetia, cioe lo Capitolo de censo, e cubo equal a numero con li altri compagni, anchor che a quel tẽpo non mi volsi scoprir con sua Eccellentia, accioche quella non tentasse de trovarli, pche sapeva che tal cosa gli saria facile p vigor della mia cosi humel piãta.*

At least some among modern scholars think this claim unfounded. See Arnaldo Masotti's introduction to the *Quesiti et Inventioni*, p. XXIII, and the same author's contribution to the Convegno di Storia delle Matematiche held in Brescia in 1959 entitled "Niccolò Tartaglia e i Suoi 'Quesiti'" (Brescia, 1962), p. 26; Moritz Cantor, *Vorlesungen über Geschichte der Mathematik* (Leipzig, 1900), II, 511.

[33] In this sense, it is incorrect to speak of del Ferro or of Tartaglia or of Cardano as the discoverer of *the* solution for *the* cubic. Though no doubt each one of them would be proud to be known as such, the fact is that, even ignoring the irreducible case, the time for synthesizing Cardano's results into a single formula had not yet arrived.

never, even through posthumous publication, enlightened the world on how far he had carried his enterprise.[34]

III

The first edition of *The Great Art* was published in 1545 by Johann Petrieus of Nürnberg. A second edition appeared twenty-five years later from the press of the Officina Henricpetrina in Basel. A third is that found in volume IV of the author's *Opera Omnia* published in Lyons by Huguetan and Ravaud in 1663.[35] The second edition is a slightly expanded version of the first but otherwise, despite a number of typographical errors that do not occur in the first, differs from it in no important respect. The third follows the text of the second, reproducing nearly all of its typographical errors and adding a number of its own.[36]

The present text is a translation of 1545 with notes to show the material added in 1570 and 1663. Minor variations in the texts are shown in footnotes in two classes of cases: (1) those in which the variation might cause trouble for a reader who wishes to compare the translation with the Latin but has only 1570 or 1663 to work from; (2) those that are otherwise of some interest, such as the substitution in 1570 and 1663 of words with a strong Greek heritage for the Latinisms of 1545—for instance, *monades* for *unitates*, *aloga* for *irrationales*, *rheta* for *rationales*, and *analoga* for *proportionales*.

In preparing this translation, I have tried to preserve a number of passages in a form rhetorically close to the original in order to give some of the feel of the original. I have not hesitated, however, to use modern terminology even in these passages—"first power" or "the unknown," for example, for the author's *res* or *positio*—and in much of the text I have used modern symbolism (including literal coefficients) even though symbols, modern or otherwise, are completely lacking in the original, unless one counts such abbreviations as *p:* (plus), *m:* (minus) and ℞ (*radix* = root) as symbols.[37] Likewise, though some may

[34] No trace of Tartaglia's projected "new algebra" was found among his papers after his death. Cantor, *op. cit.*, *supra* note 32, II, 518.

[35] In preparing this translation, I have for the most part worked from photoreproductions of the copies of 1545 and 1663 in the Columbia University Library and of 1570 in the Library of Congress.

[36] These three editions are referred to hereafter and in the footnotes that accompany the text by date alone. Where a footnote refers to "the text" the reference is to all editions in which the matter in question appears.

[37] Although our use of x, x^2, x^3, x^4, x^5, and x^6 is more economical than Cardano's *res* (or *positio* or, occasionally, *quantitas*), *quadratum*, *cubum*, *quadratum quadrati*, *primum relatum* (or *primum nomen*) and *cubum quadrati* and their abbreviations, these terms

think otherwise, I can see no advantage to translating a word like *res* as "thing" or "things." First, such a translation is likely to distract a reader from the idea that the author is trying to get across and to focus his attention on an archaism instead. Second, I have convinced myself that *res* had become, in Cardano's use of it, as much of an abstract term as is our x and that, therefore, x or "first power" is a more accurate reflection of what Cardano meant than "thing" would be.[38] Third, there are some terms which would be completely baffling if they were taken over literally but which make perfectly good sense if they are put into the modern. *Primum relatum*, for example, if put into English as "the first related" would mean nothing; translated as x^5 or "the fifth power" the intent of the author is carried into the English.

Cardano's work varies in style between the tediously prolix and the annoyingly cryptic. The latter occurs most frequently in the chapters of the book in which the author is expounding a series of working rules, particularly Chapters XXXII through XXXVI. The introductory paragraphs of these chapters read more like lecture notes than like completed expositions of their subjects and sometimes leave the reader, whether he is reading the original text or the translation, with the uncomfortable feeling that he is missing something and perhaps misunderstanding the author's meaning.

Most of the remainder of the book, on the other hand, is prolix. The least important direct cause of this is the absence of symbols. More to the point in accounting for this defect of style are two other factors. The first of these was his need, for reasons already stated, to expound each of the thirteen types of cubic equation separately just as al-Khowarizmi, some 700 years earlier, had had to expound each of the three forms of the quadratic separately. The second was undoubtedly the newness of many of the subjects Cardano was talking about and the

---

rarely pose an impediment to understanding the original text of *The Great Art*. On the other hand, the absence of anything corresponding to our parentheses, brackets, and braces often makes for trouble (cf. footnote 5, p. 50 *infra*). Compare Cardano's use of the terms *radix*, *radix universalis*, and *radix universalissima* and their abbreviations where we would use radical signs in such expressions as $\sqrt{x}$, $\sqrt{\sqrt{x} + N}$, and $\sqrt{\sqrt{\sqrt{x} + N} - \sqrt{\sqrt{x} - N}}$.

[38] Consider how baffling *res* translated as "thing" would be to a Cardano, whose acquaintance with Euclid's "Things equal to the same thing are equal to each other" was far from superficial, when confronted with his own discovery that the "thing" of a cubic or biquadratic equation may well have two or more values which are not equal to each other!

fact that he was a teacher of mathematics with a need to make himself clear in lectures to his students. The newness of his subject matter made it important for him to lead his readers through his processes by very detailed but easy stages that, given more familiarity, could have been dispensed with. The fact that he had spent years lecturing on mathematics, moreover, made this a congenial way of expounding his ideas. For the translator, to add a personal note, this prolixity is in many instances a great help for it enables him to check the correctness of his translation of the more general and abstract passages—for example, Cardano's formulations of his rules—against the problems with an ease which would otherwise be out of the question.

IIII

My indebtedness in connection with this translation is very great to two men. One is the late Professor Frederick Barry of Columbia University. It was he who, in his course on the History of Science, first made me aware of the importance of *The Great Art*, encouraged me to undertake the translation and, with great patience and without showing any signs of discouragement at what I now know was a wholly inadequate draft fit to be looked at by no one except myself, worked through its first few chapters with me. The second is Dirk J. Struik, Professor Emeritus of Mathematics at the Massachusetts Institute of Technology. He was kind enough to agree to go over the almost final version of the translation when I, a total stranger, wrote to him asking if he would be willing to do so. Not only did he do this for me but he also volunteered to act, in effect, as a literary agent and found the book its publisher. I am also obligated to Professor Morris Kline of New York University for encouraging publication of the translation and to the staff of the M.I.T. Press for their painstaking work in preparing the copy for the printers and seeing it through the press.

T. R. W.

Alexandria, Virginia
February 1968

THE GREAT ART[1]

or

The Rules of Algebra

by

GIROLAMO CARDANO

Outstanding Mathematician, Philosopher and Physician
In One Book, Being the Tenth in Order of the Whole
Work on Arithmetic Which is called the Perfect Work

In this book, learned reader, you have the rules of algebra (in Italian, the rules of the coss[2]). It is so replete with new discoveries and demonstrations by the author — more than seventy of them — that its forerunners [are] of little account or, in the vernacular, are washed out. It unties the knot not only where one term is equal to another or two to one but also where two are equal to two or three to one. [3]It is a pleasure, therefore, to publish this book separately so that, this most abstruse and unsurpassed treasury of the entire [subject of] arithmetic being brought to light and, as in a theater, exposed[4] to the sight of all, its readers may be encouraged and will all the more readily embrace and with the less aversion study thoroughly the remaining books of the Perfect Work which will be published volume by volume.[3]

[1] The original title page was omitted in 1663.

[2] "Coss" and its variants were familiar terms for algebra in fifteenth-century and sixteenth-century Italian, German, and English. See David Eugene Smith, *History of Mathematics* (Boston, 1923–1925), I, 320, 328; II, 392.

[3] In lieu of this passage 1570 has the following:
It is a pleasure, therefore, to publish this book anew partly so that, this most abstruse and clearly unsurpassed treasury of the entire [subject of] arithmetic being brought to light and, as in a theater, exposed[4] to the sight of all, its readers may be encouraged and will all the more readily embrace and with the less aversion study thoroughly the remaining books of the Perfect Work and partly because it has recently been revised and enlarged by the author.

[4] 1545 has *exposito*; 1570 has *opposito*.

[DEDICATION]

Girolamo Cardano, Physician, to the most erudite *Andreas Osiander*,[1] greetings:

I have considered nothing so deeply, learned Andreas, as the names of those who, by their writings, deserve to be commended to posterity. I have especially asked whether they combine liberal learning with erudition. Hence, since I know that you have a far from mediocre knowledge not only of Hebrew, Greek, and Latin letters but also of mathematics and since, moreover, [you are] the most humane man I have ever encountered, it seemed to me that this, my book, could be better dedicated to no one than you by whom it may be corrected (if my pen has gone further than the power of my mind), read with pleasure, understood, and, indeed, recommended authoritatively. Unless I am mistaken, others will follow this example and dedicate their works [to you], whatever branch of learning they pursue. Accept this, therefore, as a perpetual testimonial of my regard for you, of your assistance to me, and, above all, of your distinguished scholarship. And, although you may well be such that your merits are known to all men, yet just as Alexander and Caesar wished to have their most notable deeds inscribed in the chronicles of others and as Plato, who preserved such marvels in his own writings, still desired to be praised in the writings of others, so I hope that this, my offering to you, will not be displeasing, whatever its merit, because in such matters there is a certain fortune which controls and the better perish while the poorer survive. And whatever your judgment may be in matters of this sort, it nevertheless is certain to me that it is my duty to repay my debts. I hope also that, by this clear example, there may be known my affection for all who have that same candor of spirit for which you are recognized among the scholars of our time. But perhaps a better occasion [for this] will be given; if not, I still do not wish this one, such as it is, to be lost to me.

Farewell. Pavia, the 5th of the Ides of January, 1545.

[1] Andreas Osiander (1498–1552) of Nürnberg was one of the minor leaders of the Reformation. He is generally credited with authorship of the anonymous preface to Copernicus' *De Revolutionibus* (see Preserved Smith, *A History of Modern Culture* (New York, 1930), I, 40 ff; Henry Osborn Taylor, *Thought and Expression in the Sixteenth Century* (2d ed., New York, 1930), II, 336–337; Thomas S. Kuhn, *The Copernican Revolution* (Modern Library ed., New York, 1959), p. 187; A. C. Crombie, *Medieval and Early Modern Science* (New York, 1959), II, 168, 186). James Eckstein in his *Jerome Cardan* (Baltimore, 1946), pp. 7, 23, credits Osiander with having edited *The Great Art.*

Index to the Matters Contained in this Book[1]

Chapter

[1] Some of the chapter titles in this Index vary in minor respects from those that appear at the heads of the chapters themselves.

[2] *De aestimatione generali & equatione, cum media denominatio aequatur extremae & numero.*

[3] *De regulis maioribus singularibus.*

CHAPTER I

On Double Solutions in Certain Types of Cases

1. This art originated with Mahomet the son of Moses the Arab.[1] Leonardo of Pisa is a trustworthy source for this statement.[2] There remain, moreover, four propositions of his with their demonstrations, which we will ascribe to him in their proper places.[3] After a long time, three derivative propositions were added to these. They are of uncertain

[1] Better known today as al-Khowarizmi; died *c.* A.D. 840. His *Algebra* is most readily available in L. C. Karpinski's edition entitled *Robert of Chester's Latin Translation of the Algebra of al-Khowarizmi* (New York, 1915), which also contains an English translation. See also Guillaume Libri's *Histoire des Sciences Mathématiques en Italie* (Paris, 1838), I, 261 ff., where a somewhat different Latin version, presumably by Gerard of Cremona (*cf.* Karpinski, *ibid.*, p. 82) is set out, and Baldassarre Boncompagni's *Della Vita e delle Opere di Gherardo Cremonese* (Atti dell' Academia Pontificia de' Nuovi Lincei, vol. IV; Rome, 1851), where there is still another Latin version which Boncompagni attributes to Gerard of Cremona but which Karpinski believes (*ibid.*, p. 42) is not his, though probably based on it. I regret that I have not seen a fourth version, an English translation by Frederic Rosen direct from the Arabic, entitled *The Algebra of Mohammed ben Musa* (London, 1831).

[2] Better known today as Leonardo Fibonacci; *fl.* 1202. In his *Liber Abbaci* (1202), as printed in volume I of Baldassare Boncompagni's edition of his *Scritti* (Rome, 1857), the simple notation "Maumeht" appears in the margin at the beginning of Part III, entitled *De Solutione Quarundam Quaestionum Secundum Modum Algebra et Almuchabale* (p. 406). Whether the manuscript that Cardano worked from had more of an attribution than this cannot be said.

In the 39th problem in his *Ars Magna Arithmeticae* (*Opera Omnia*, IV, 374), Cardano says that he first found that "the name of the author of the book which is called Algebra" was Mahomet in a large paper volume "on the first desk in the library of San Antonio in Venice." The name of the author, he says, had been erased, but the book was dated 1202. The book also contained marginal notations in the hand of Luca Paccioli and so much of its contents are to be found in Luca's book, Cardano adds, that it is clear that he transcribed the whole of it.

[3] If Cardano meant to include al-Khowarizmi's three simple cases ($x^2 = N$; $x^2 = ax$; $ax = N$), there are six propositions surviving; if he meant to include only those for the three-term quadratic cases, there are not four survivors. There are, however, four demonstrations — two for $x^2 + ax = N$ and one each for $x^2 = ax + N$ and $x^2 + N = ax$. Cardano later forgets to mention the authorship of the two he actually uses (see pp. 33 and 35).

authorship, though they were placed with the principal ones by Luca Paccioli.[4] I have also seen another three, likewise derived from the first, which were discovered by some unknown person. Notwithstanding the latter are much less well known than the others, they are really more useful, since they deal with the solution of [equations containing] a cube, a constant, and the cube of a square.

In our own days Scipione del Ferro of Bologna[5] has solved the case of the cube and first power equal to a constant, a very elegant and admirable accomplishment. Since this art surpasses all human subtlety and the perspicuity of mortal talent and is a truly celestial gift and a very clear test of the capacity of men's minds, whoever applies himself to it will believe that there is nothing that he cannot understand. In emulation of him, my friend Niccolò Tartaglia of Brescia,[6] wanting not to be outdone, solved the same case when he got into a contest with his [Scipione's] pupil, Antonio Maria Fior,[7] and, moved by my many entreaties, gave it to me.[8] For I had been deceived by the words of Luca Paccioli, who denied that any more general rule could be discovered than his own.[9] Notwithstanding the many things which I

[4] Also known as Luca di Burgo; *fl.* 1494. His *Sũma de Arithmetica Geometria Proportioni & Proportionalita* was first published in Venice in 1494. On p. 149r he discusses the cases of $x^4 + N = bx^2$, $x^4 + bx^2 = N$, and $x^4 = bx^2 + N$, which are presumably those to which Cardano here refers.

[5] Born *c.* 1465; died 1526.

[6] Born *c.* 1506; died 1557.

[7] *Fl. c.* 1515.

[8] Cardano repeats this statement at the beginning of Chapter XI, p. 96 *infra.* See also Chapter XXVIII of the *Ars Magna Arithmeticae* (*Opera Omnia*, IV, 341), the title of which is "De Capitulo Generali Cubi et Rerum Aequalium Numero, Magistri Nicolai Tartaglia, Brixiensis," and the passage from the *De Libris Propriis* quoted *infra* p. 239. These repeated attributions plus Cardano's acknowledgements of indebtedness to Ferrari make clear that, regardless of the question whether Cardano was justified in publishing what, according to Tartaglia, he obtained from him under a promise of strict secrecy, such a writer as Herbert W. Turnbull far overshoots the mark when he says, in his *The Great Mathematicians* (4th ed., 1951, reprinted in James R. Newman, *The World of Mathematics* [New York, 1956]), I, 119: "GIROLAMO CARDANO (1501–1576) was a turbulent man of genius, very unscrupulous, very indiscreet, but of commanding mathematical ability. . . . He was interested one day to find that Tartaglia held a solution of the cubic equation. Cardan begged to be told the details, and eventually under a pledge of secrecy obtained what he wanted. Then he calmly proceeded to publish it as his own unaided work in the *Ars Magna* which appeared in 1545. . . . He seems to have been equally ungenerous in the treatment of his pupil Ferrari, who was the first to solve a quartic equation. Yet Cardan combined piracy with a measure of honest toil, and he had enough mathematical genius in him to profit by these spoils. . . ."

[9] Cardano is probably referring to Paccioli's characterization of $x^4 + bx^2 = ax$ and $x^4 + ax = bx^2$ as *impossibile*; see Paccioli's *Sũma*, p. 149r.

had already discovered, as is well known, I had despaired and had not attempted to look any further. Then, however, having received Tartaglia's solution and seeking for the proof of it, I came to understand that there were a great many other things that could also be had. Pursuing this thought and with increased confidence, I discovered these others, partly by myself and partly through Lodovico Ferrari,[10] formerly my pupil. Hereinafter those things which have been discovered by others have their names attached to them; those to which no name is attached are mine. The demonstrations, except for the three by Mahomet[11] and the two by Lodovico, are all mine. Each is individually set out under a proper heading and, following the rule, an illustration is added.

Although a long series of rules[12] might be added and a long discourse given about them, we conclude our detailed consideration with the cubic, others being merely mentioned, even if generally, in passing. For as *positio* [the first power] refers to a line, *quadratum* [the square] to a surface, and *cubum* [the cube] to a solid body, it would be very foolish for us to go beyond this point. Nature does not permit it. Thus, it will be seen, all those matters up to and including the cubic are fully demonstrated, but the others which we will add, either by necessity or out of curiosity, we do not go beyond barely setting out. In everything, however, the worth of the preceding books [of this work], especially the third and fourth books, should be kept in mind, lest I be thought trifling when I repeat or obscure when I skip over something.

2. Now let it be remembered that we have shown that there are odd powers and even powers. The square, the square of the square, the cube of the square, and so on always skipping one, are even, while the first power,[13] the cube, and the fifth and seventh powers[14] we call odd. [It will be remembered also that] 9 is derivable equally from 3 and -3, since a minus times a minus produces a plus. But in the case of

[10] Born 1522; died 1560. Our best source of information about him is the *Vita* by Cardano (*Opera Omnia*, IX, 568–569), where he described him thus: *Fuit parvae staturae, iucundi vultus, blando sermone, prudens in rebus exigui momenti, comptus, naso parvo, non tamen deformi, roseo colore. Verbo; undequaque ad decorem formatus.* The conclusion of the *Vita* reads: *Haec fuit infoelix vita Ludovici Ferrarij, ingenio et eruditione in Mathematicis nulli Secundi, sed in humanis rebus minime sapientis, et in Deum parum pij, et Divos omnes ex consuetudine turpiter execraretur, adeoque violenta ira processerat, ut ego raro ad illum accederem, nec alloqui auderem.*

[11] See note 3 *supra*.

[12] 1545 has *innumerata capitulorum series*; 1570 and 1663 have *longa capitulorum series*.

[13] *rem seu positionem*.

[14] 1545 has *primum ac secundum nomen*; 1570 and 1663 have *primum ac secundum relatum*.

the odd powers, each keeps its own nature: [15]it is not a plus unless it derives from a true number, and a cube whose value is minus,[15] or what we call *debitum*, cannot be produced by any expansion of a true number. It behooves us to remember this very clearly.

3. If, therefore, an even power is equal to a number, its root has two values, one plus, the other minus, which are equal to each other. As, if

$$x^2 = 9,$$

x is 3 or -3. And if

$$x^2 = 16,$$

x is 4 or -4. And if

$$x^4 = 81,$$

x equals 3 or -3. To set out other even powers is not necessary, since [even] x^4 belongs to the derivative cases.

Now if you will pay strict attention to what I am saying you can satisfy a wish of yours with this rule. For if the square and the square of a square are equated to a number, the result will be the same as in the simple case; that is, there are two solutions, one plus and one minus, which are equal to each other.[16] Thus, if

$$x^4 + 3x^2 = 28,$$

x equals[17] 2 or -2.

And, again, if the square of a square and a number are equal to a square, we will readily demonstrate in Chapter VIII that there are two true solutions and that there are, at the same time, corresponding and equal negative solutions.[18] Thus, if I say

$$x^4 + 12 = 7x^2,$$

x equals 2, -2, $\sqrt{3}$, and $-\sqrt{3}$, and thus there are four solutions. But if a true solution is lacking, a negative one will also be lacking. Thus, since there is no true solution for

$$x^4 + 12 = 6x^2,$$

[15] 1663 omits this part of the sentence.

[16] I.e., if $x^4 + ax^2 = N$, $x = +s, -s$.

[17] 1545 has *est*; 1570 and 1663 have *valet*.

[18] I.e., if $x^4 + N = ax^2$, $x = +r, +s, -r, -s$.

a fictitious one (for such we call that which is *debitum* or negative) is also lacking.

If the square of a square is equal to a number and a square, there is always one true solution and another and fictitious solution equal to it.[19] Thus, in

$$x^4 = 2x^2 + 8,^{[20]}$$

x equals 2 or -2. The same reasoning holds for all other even powers when they are joined with a number. How this comes about by depression, we have shown fully in the fourth book.

4. For an odd power, there is only one true solution and no fictitious one when it is equated to a number alone. Thus, if

$$2x = 16,$$

x equals 8, and if

$$2x^3 = 16,$$

x equals 2. It is always presumed in this case, of course, that the number to which the power is equated is true and not fictitious. To doubt this would be as silly as to doubt the fundamental rule itself for, though opposite reasoning must be observed[21] in opposite cases, the reasoning is still the same.

Where several [odd] powers, or even a thousand of them, are compared with a number, there will be one true solution and none that is fictitious. Thus, if

$$x^3 + 6x = 20,$$

there is no other solution than 2, either true or false.

5. When two powers and a constant are compared, either both powers are odd and the comparison is made either with the highest or with the middle power (for we have dealt in the preceding rule with the case where it is made with the number) or one is odd and the other is even (for we have already treated in the third rule with the case where both are even).

If, therefore, the highest power (namely, a cube) and constant are equated to the middle power (that is, the first power), see whether multiplying two-thirds the coefficient of the first power by the square root of one-third the same yields the constant or something greater or

[19] I.e., if $x^4 = ax^2 + N$, $x = +s, -s$.

[20] 1570 and 1663 have 80.

[21] 1545 has *persequenda*; 1570 and 1663 have *observanda*.

less. If the product and the constant are precisely[22] the same, the unknown has two values, and one of the two is true, namely the same square root which has been multiplied.[23] For example,

$$x^3 + 16 = 12x.$$

Multiply 8, which is two-thirds of 12, the coefficient of x, by 2, the root of 4, which is one-third the coefficient of x. This produces 16, the constant of the equation. Therefore the solution is 2, the root of 4.

The other solution is false and corresponds to the true solution [for the case] of the cube equal to the same constant and the first power with the same coefficient.[24] For example, if

$$x^3 = 12x + 16,$$

the true solution is 4. Therefore, if

$$x^3 + 16 = 12x,$$

the value of x is -4, since $12x$ is -48 and the cube of -4 is -64, to which 16 is to be added, making -48.

But if the product of two-thirds the coefficient of x and the square root of one-third the same is greater than the constant, there will be three solutions, two that are true and a third that is false. For example, in

$$x^3 + 9 = 12x,$$

the true solutions are 3 and $\sqrt{5\frac{1}{4}} - 1\frac{1}{2}$, and the third or false one is the sum of these two and corresponds to the true solution for the cube equal to the same first power and the same constant.[25] This is $\sqrt{5\frac{1}{4}} + 1\frac{1}{2}$.[26] Thus the remaining fictitious solution of which we spoke in the other example is the sum of two true ones, but since the true ones are equal to each other, the fictitious one is twice the true one.[27]

[22] 1545 has *praecise*; 1570 and 1663 have *ad unguem*. The same variance occurs elsewhere in the book but is not further noted.

[23] I.e., if $x^3 + N = ax$, and if $\frac{2}{3}a\sqrt{\frac{1}{3}a} = N$, then $x = r, -s$, and $r = \sqrt{\frac{1}{3}a}$.

[24] I.e., if $x^3 + N = ax$, and if $\frac{2}{3}a\sqrt{\frac{1}{3}a} = N$, then $x = -s$, and $-s$ equals the $+s$ of $x^3 = ax + N$.

[25] I.e., if $x^3 + N = ax$, and if $\frac{2}{3}a\sqrt{\frac{1}{3}a} > N$, then $x = +r, +s, -t$, and $r + s = t$, and $-t =$ the $+t$ in $x^3 = ax + N$.

[26] This means that the fictitious solution in $x^3 + 9 = 12x$ is $-\sqrt{5\frac{1}{4}} - 1\frac{1}{2}$.

[27] *et ita reliqua ficta, de qua diximus, in alio exemplo, aggregatur ex duabus veris, sed quia verae sunt invicem aequales, ideo ficta semper dupla est verae.* With the translation of this passage given above and of the further passage set out below at note 30, compare

It is clear, therefore, that the false or fictitious solutions in the case of the cube and a number equal to the first power correspond to the true solutions in the case of the cube equal to the first power and constant if [the coefficient of] the first power and the constant are the same.

If the product of the square root of one-third the coefficient of x and two-thirds the same is less than the given constant, there will be no true solution but a fictitious one equal to the true one in the case of the cube equal to the same number of x's and the same constant.[28] For instance, although

$$x^3 + 21 = 2x$$

lacks a true solution, yet it has a fictitious one, -3, and this is the true solution for

$$x^3 = 2x + 21.$$

6. It is not difficult from this to discover how many solutions there are for the case of the cube equal to the first power and constant. If the product of two-thirds the coefficient of x and the square root of one-third the same is equal to the constant, there are two solutions — a true one equal to the fictitious one of the preceding rule, and a fictitious

---

J. E. Montucla, *Histoire des Mathématiques* (Paris, 1960, reprint of 1799–1802 edition), I, 594–595. After pointing out that "Cardano was the first to perceive the multiplicity of values for the unknown in equations and the distinction between positives and negatives," he adds: "Yet let us notice, in order not to give Cardano too much credit, that his discovery was not fully developed: not only did he say nothing about the use of these negative roots, which he probably regarded as useless, but he was also mistaken with respect to equations which have several equal roots affected by the same sign. . . . This error, however, is excusable in a time when algebra was used only for the solution of numerical problems. For let us suppose a problem of this nature which leads to the last of the equations above [$x^3 + 16 = 12x$; $x = +2, +2, -4$]; who would have been the analyst who would have noticed that this gave 2 twice and -4? He could not have helped but regard the two solutions as the same, not being able to distinguish one from the other. Simple arithmetic would have thrown no light on the subject and it is only the application of algebra to curves which can make one understand the distinction of which we speak."

On the other hand, Moritz Cantor in his *Vorlesungen über Geschichte der Mathematik* (Leipzig, 1900), II, 505, favors the reading given in the text above and asks whether (particularly in view of Cardano's back reference to this chapter in Chapter XVIII) there can be any doubt that Cardano had at least a vague idea of the existence of equal roots with the same sign, and Guglielmo Libri, in his *Histoire des Sciences Mathématiques en Italie* (Paris, 1838), III, 171, cites the present passage as evidence that Cardano "knew and treated with equal roots." In addition, the demonstration and rule set out in Chapter XIII of *The Great Art* clearly lead to equal roots with the same sign in certain cases.

[28] I.e., if $x^3 + N = ax$, and if $\frac{2}{3}a\sqrt{\frac{1}{3}a} < N$, then $x = -s$, and $-s = +s$ in $x^3 = ax + N$.

one equal to the true.[29] Hence the true one is twice the fictitious just as, in that instance, the fictitious is twice the true.[30] For example, in

$$x^3 = 12x + 16,$$

the true solution is 4 and the fictitious one is -2, since if

$$x^3 + 16 = 12x$$

the true solution is 2 and the fictitious one -4.

But if from the aforesaid multiplication there comes a result greater than the constant of the equation, there will be one true solution corresponding to the false one under the preceding rule, and two false ones corresponding to the true ones of the preceding rule.[31] If, for instance,

$$x^3 = 12x + 9,$$

the two false solutions are $-(\sqrt{5\frac{1}{4}} - 1\frac{1}{2})$[32] and -3, and the true one is $\sqrt{5\frac{1}{4}} + 1\frac{1}{2}$, and thus you see how the false [in this case] corresponds to the true [in the preceding case] and the true to the false in turn, and moreover how the two false ones give rise to the true, for the sum of $\sqrt{5\frac{1}{4}} - 1\frac{1}{2}$ and 3 is $\sqrt{5\frac{1}{4}} + 1\frac{1}{2}$.

But if this multiplication yields something less than the constant, there is only one solution — a true one — just as under the preceding rule there is only one solution — a fictitious one.[33] For instance, if

$$x^3 = 2x + 21,$$

x will be 3, just as in the case of

$$x^3 + 21 = 2x,$$

the fictitious solution is -3.

7. In those cases in which a constant, an even power, and an odd power are equated in turn, either the highest power is even (as when the square, the first power, and a constant are equated) or the highest power is odd (as when a cube and a square are equated to a constant.)

[29] I.e., if $x^3 = ax + N$, and if $\frac{2}{3}a\sqrt{\frac{1}{3}a} = N$, then $x = +r, -s$ which correspond, respectively, to $-r, +s$ in $x^3 + N = ax$.

[30] *ideo vera est dupla fictae, quia ibidem ficta est dupla verae.* See footnote 27 *supra.*

[31] I.e., if $x^3 = ax + N$, and if $\frac{2}{3}a\sqrt{\frac{1}{3}a} > N$, $x = +r, -s, -t$, and $+r, -s, -t$ correspond respectively to $-r, +s, +t$ in $x^3 + N = ax$.

[32] Cardano expresses this thus: "R̸ 5¼ *m:* 1½ *m:*".

[33] I.e., if $x^3 = ax + N$, and if $\frac{2}{3}a\sqrt{\frac{1}{3}a} < N$, then $x = +r$, and $+r = -r$ in $x^3 + N = ax$.

If, then, a square is equated to the first power and a constant, it will have two solutions, a true one equal to the fictitious one in the case of the square and first power equal to the same constant, and a fictitious one equal to the true one in the other case.[34] For example, if

$$x^2 + 4x = 21,$$

the true solution is 3 and the fictitious one -7. And if

$$x^2 = 4x + 21,$$

the true solution is 7 and the fictitious one -3. Hence the true solutions having been obtained, the fictitious are also known, as under the preceding rule but in a different manner, for here extreme is compared to extremes, there the mean was compared to the extremes. There the case of the cube and constant equal to the first power was compared to the case of the cube equal to the first power and constant; here the case of the square and first power equal to a constant is compared to the case of the square equal to the first power and constant.

But when the square and constant are equated to the first power, and the case is possible, then there are two true solutions. Say, for instance, that

$$x^2 + 12 = 7x.$$

The unknown may then be 4 or 3, both of which can be verified. When, however, the constant is equal to the square of one-half the coefficient of the first power,[35] there is only one solution, viz., one-half the coefficient of the first power.[36] In this case, moreover, there can never be a fictitious or negative solution, but [except in the instance just cited] where there is one true solution there are two of them and where there is no true solution there still cannot be a fictitious one.

8. If, now, a solution is sought for cases of the cube, square, and constant, then if the cube is equal to the square and constant, there is only one true solution.[37] For example, in

$$x^3 = 3x^2 + 16,$$

x equals 4, and no other solution can be discovered.

[34] I.e., if $x^2 = ax + N$, then $x = +r, -s$, and $+r, -s = -r, +s$ in $x^2 + ax = N$.

[35] Here and in its next occurrence, the text has *radicum* instead of *rerum*.

[36] I.e., if $x^2 + N = ax$, then $x = +r, +s$, unless $(a/2)^2 = N$, in which case $x = +r$, and $r = a/2$.

[37] I.e., if $x^3 = ax^2 + N$, $x = +s$.

Note. In all cases in which there is only one solution, it is easily and clearly discovered, as in the case of the cube and first power equal to a constant or of the cube equal to the square and constant, and in the case of the cube equal to the first power and constant where two-thirds the coefficient times the square root of one-third the same is less than the constant. The same is true where the cube and constant are equated to the first power, and it is impossible to have anything except a fictitious solution. The other cases, however, in which the solution is multiplex are more difficult and confusing.

If, therefore, the cube and the square are equal to a constant and if the product of one-third the coefficient of x^2 and the square of two-thirds the same is less than the constant of the equation, x has only one value and it is positive. This value is the same as the fictitious solution in the corresponding case of the cube and constant equal to the square with the same coefficient.[38] For example,

$$x^3 + 3x^2 = 20.$$

Now, since the product of one-third the coefficient of x^2 and 4, the square of two-thirds the same, is less than 20, I say that there is only one solution and the value of x is 2 and this, taken in the negative, is the solution for

$$x^3 + 20 = 3x^2.$$

If the product of this multiplication is the same as the constant, there will be one true and two fictitious solutions, and the true one will correspond to the fictitious one in the other case, and the fictitious ones will correspond to the true.[39] For example, if

$$x^3 + 11x^2 = 72,$$

x is $\sqrt{40} - 4$ for the true value and the false values are -3 or $-(\sqrt{40} + 4)$. And if

$$x^3 + 72 = 11x^2,$$

the true values are 3 and $\sqrt{40} + 4$ and the false one is $-(\sqrt{40} - 4)$. Hence by finding a fictitious one, we also find a corresponding true value for the other case.

[38] I.e., if $x^3 + ax^2 = N$, and if $\frac{1}{3}a(\frac{2}{3}a)^2 < N$, then $x = +s$, and $+s = -s$ in $x^3 + N = ax^2$.

[39] I.e., if $x^3 + ax^2 = N$, and if $\frac{1}{3}a(\frac{2}{3}a)^2 = N$, then $x = +r, -s, -t$, and $+r, -s, -t = -r, +s, +t$ in $x^3 + N = ax^2$.

Note. It will be noted from this that certain cases have two values and others have one, and when there are three in one part of a case there will be only one in another, as in the case of the cube equal to the first power and a constant in the smaller part [i.e., in instances where the product of two-thirds the coefficient of x and the square root of one-third the same is less than the constant] and the case of the cube and square equal to the constant and the case of the cube and constant equal to the square or first power, for in one part there are three solutions, in the other only one. Similarly, in the case of the fourth power and constant equal to the square: in one part there are four solutions, in another none. There are those which have two solutions throughout, as the case of the square and first power equal to a constant or the case of the square equal to the first power and constant. And there are those that have only one solution, as the case of the cube and first power equal to a constant and [= whereas] the case of the square and constant equal to the first power has two solutions in one instance and none in the other.

Note.[40] You know, furthermore, that the solutions for the cases of the cube and square equal to a constant and of the cube and constant equal to the square are such that the difference between the true and the fictitious solutions is always the coefficient of the second power. Thus, if

$$x^3 + 72 = 11x^2,$$

the false solution is $\sqrt{40} - 4$[41] and the true ones are $\sqrt{40} + 4$ and 3. The difference between $\sqrt{40} - 4$ and $7 + \sqrt{40}$ is 11, the coefficient of the second power. So, also, if

$$x^3 + 11x^2 = 72.$$

9. In those cases, however, in which there are two odd powers, an even power, and a constant, if the cube and first power are equal to the square and constant, there may be three solutions, all true and none fictitious,[42] since, as has been said, the cube of a minus quantity is a minus, and thus [if there were a negative solution] a minus would equal a plus, which cannot be.

Where the cube, square, and first power are equal to a constant, there will be three solutions, one positive, two negative,[43] and if, with

[40] 1570 and 1663 omit this word.
[41] I.e., this solution is $-(\sqrt{40} - 4)$ or $4 - \sqrt{40}$.
[42] I.e., if $x^3 + ax = bx^2 + N$, $x = +r, +s, +t$.
[43] I.e., if $x^3 + bx^2 + ax = N$, $x = +r, -s, -t$.

the same coefficients, the square can be equated to the first power, constant, and cube, what are here true solutions will be the fictitious ones in the other instance.[44] For example,

$$x^3 + 6x^2 + 3x = 18.^{45}$$

Now the true solution can be had from this case and the false ones from the case of

$$x^3 + 3x + 18 = 6x^2,$$

and one of them is 3 and the other $\sqrt{8\frac{1}{4}} + 1\frac{1}{2}$. Hence -3 or $-(\sqrt{8\frac{1}{4}} + 1\frac{1}{2})$[46] is the fictitious solution for

$$x^3 + 6x^2 + 3x = 18$$

and with this there is also a third and true value.

From this may be had the three solutions for the case of the cube, first power, and constant equal to the square. (If any solution is possible, it is known from the proper rules.) Two of these are true and equal, as has been said, to the [negative] solutions for the case of the cube and the same number of squares and first powers equal to the same constant, as in the preceding example. The third, however, is fictitious and corresponds to the true solution for the other case.[47] Hence the solution for the case of

$$x^3 + 6x^2 + 3x\ [= 18]$$

is the same, but negative, as the true solution in the case of

$$x^3 + 3x + 18 = 6x^2.$$

But if the number of squares is too few for them to be equated to the cube, the first power, and constant [with a positive value for x], then there is one true solution and none false [48]in the case of the cube, square, and first power equal to the constant,[48] and there is one false solution and none true in the case of the square equal to the cube, first power, and constant. For instance, in

$$x^3 + x^2 + 2x = 16,$$

[44] I.e., if $bx^2 = x^3 + ax + N$, $x = +s, +t$, and $+s, +t$ correspond, respectively, to $-s, -t$ in $x^3 + bx^2 + ax = N$.

[45] 1663 and, perhaps, 1570 have 8.

[46] In this case Cardano writes it as "*m*: ℞ $8\frac{1}{4}$ *p*: $1\frac{1}{2}$."

[47] I.e., if $x^3 + ax + N = bx^2$, $x = +r, +s, -t$, and $+r, +s, -t$ correspond, respectively to $-r, -s, +t$ in $x^3 + bx^2 + ax = N$.

[48] 1663 omits this phrase.

x equals 2, and this is the fictitious solution for

$$x^3 + 2x + 16 = x^2.$$

It is clear, therefore, that the case of the cube, second power, and first power equal to a constant and that of the cube, first power, and constant equal to the square correspond to each other.

10. Similarly, the case of the cube equal to the square, first power, and constant corresponds to that of the cube, square, and constant equal to the first power and, therefore, if [in the latter] the coefficient of the first power is too small, there is a single fictitious solution which is equal to the corresponding true one in the other case of the cube equal to the same number of squares and first powers and the same constant. For example, if

$$x^3 = 2x^2 + x + 6,$$

x equals 3, neither more nor less, since if

$$x^3 + 2x^2 + 6 = x,$$

there can be no true solution, but the fictitious one will be -3, which is the true one in the other case.

But if the first power is such that the case of the cube, square, and constant equal to the first power can have a true solution, then it will have two such and there will [also] be a fictitious one, these corresponding to the two fictitious solutions and one true solution of the other case. For example, if

$$x^3 + 3x^2 + 6 = 20x,$$

there will be two true solutions, namely 3 and $\sqrt{11} - 3$, and a fictitious one, namely $-(\sqrt{11} + 3)$. Hence, the true value of

$$x^3 = 3x^2 + 20x + 6$$

is $\sqrt{11} + 3$ and the two fictitious ones are -3 and $-(\sqrt{11} - 3)$.

11. For the same reason, the cases of the cube and square equal to the first power and constant and of the cube and constant equal to the square and first power correspond to each other. If a case of the cube and constant equal to the first power and square does not have a true solution, it will still have a false one equal to the true one of the other case. For example,

$$x^3 + 72 = 6x^2 + 3x.$$

The fictitious value of x is -3, and this is the true solution for

$$x^3 + 6x^2 \text{[49]} = 3x + 72.$$

And, just as the case of

$$x^3 + 72 = 6x^2 + 3x$$

lacks a true solution, so the case of

$$x^3 + 6x^2 = 3x + 72$$

lacks a false one.

But where a case of the cube and constant equal to the square and first power has a true solution, it will have two of them and a false one in addition, corresponding to the two false solutions and one true solution in the other case. For example,

$$x^3 + 4 = 3x^2 + 5x.$$

The true solutions are 4 and $\sqrt{1\frac{1}{4}} - \frac{1}{2}$ and the false one is $-(\sqrt{1\frac{1}{4}} + \frac{1}{2})$, and the latter is the true solution for the case of

$$x^3 + 3x^2 = 5x + 4$$

and the other two, namely 4 and $\sqrt{1\frac{1}{4}} - \frac{1}{2}$ become negative and fictitious in the same case.

12. It is also manifest that if the fourth power, the first power, and a constant are compared with each other, the seventh rule will fit them to a T, as it does the case of the square, first power, and constant. By bringing one case to bear on another, the same reasoning [may be applied] to the other derivatives.

DEMONSTRATION

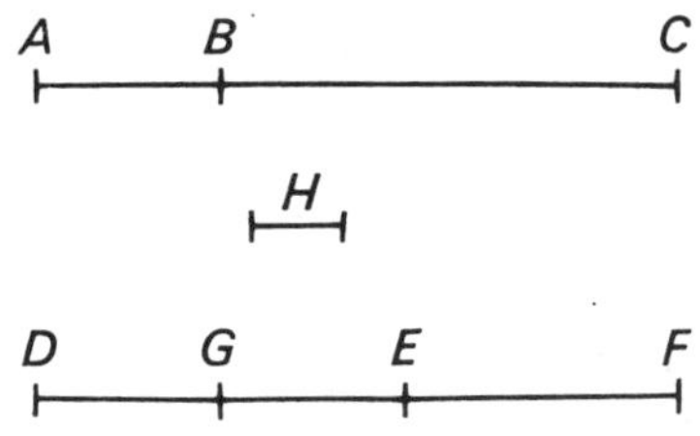

13. Now it is meet that we should show these very wonderful things by a demonstration, as we will do thoughout this whole book, so that, beyond mere experimental knowledge, reasoning may reinforce belief in them. Let, therefore, for example, AB, the cube, and BC, the constant, equal DE, the second power, and EF, the first power, and let H be a true solution. Since, therefore, by supposition, AC is equal to DF, draw DG equal to AB. Now since DE

[49] 1570 and 1663 have $9x^2$.

is greater than AB by GE, and BC is equal to GF by common consent, BC will be greater than FE by GE, and BC is greater than EF by as much as DE is greater than AB.

Now assume H to be negative and a fictitious solution. Then AB and EF will be negative, but DE and BC will remain positive. Since, therefore, the difference between AB and DE is GE and the difference between BC and EF is also GE, subtracting AB from DE and EF from BC yields as much as adding them negatively would. If follows, therefore, assuming the value of x to be $-H$, that

$$AB + DE = BC + EF,$$

for the sum of either is the remainder GE. Hence the cube and second power can be equated to the first power and constant by the same method, and the value of x is $-H$, that is the same value in one case as in the other.

It follows, also, that the sum of the parts in one case is equal to their difference in the other. As, if I say,

$$x^3 + 10 = 6x^2 + 8x$$

and the value of x in this case is true, then in the case of

$$x^3 + 6x^2 = 8x + 10$$

it will be fictitious, for the sum of x^3 and $6x^2$[50] [when x is a fictitious number] is equal to the difference between x^3 and $6x^2$[50] when it is true, and [the difference between] 10 and $8x$ will, given a truc solution, be the same as [the sum of] $8x$ and the same number when the solution is fictitious.[51]

[50] In these two instances, Cardano switches from the usual *quadratum* to *census* for x^2.

[51] At this point, 1570 and 1663 add a new paragraph reading thus:

14. When a constant and the highest power are equal to one or more middle powers, there are two solutions for x since, whatever their coefficients, the middle terms can exceed the extreme terms, as $100x^2$ can be greater than x^3. Now let AB be the value of x. The cube can then be made to equal $100x^2$ either by decreasing the estimated value of x and allowing the given number to remain the same (in which case [the value of x becomes] AC) or by increasing the cube and thus increasing the value of x (in which case it becomes AD). Therefore

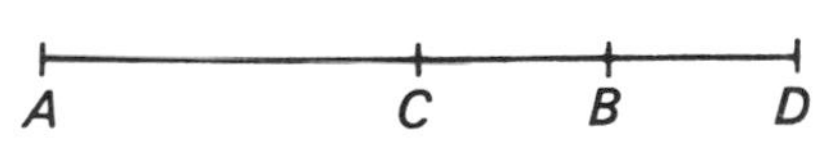

$$100x^{2}\,{}^{52} = x^3 + 1$$

has two solutions. Similarly, if there are many middle terms, even a hundred of them, since they follow the reasoning [applicable to] one, for as the value of the unknown changes all the middle powers likewise increase or decrease. But if the extreme powers and the middle powers, taken alternately, are equated to each other (as if the cube and first power are equated to the square and a constant), I say that there may be three solutions. Thus, let A, the constant, and B, the number of squares, equal K, the cube, and F, the number of x's, and let x be DE. If F is assumed to be large and DE small, an equation can be made, since the squares and the cube become smaller because of the smallness of DE. But if the squares exceed the cube and if the x's are placed beside the latter, as said, there will be two solutions — either one that is increased because of the magnitude of x^3 or one that is diminished by the increase in the x's. Therefore, there will be three solutions [in all].

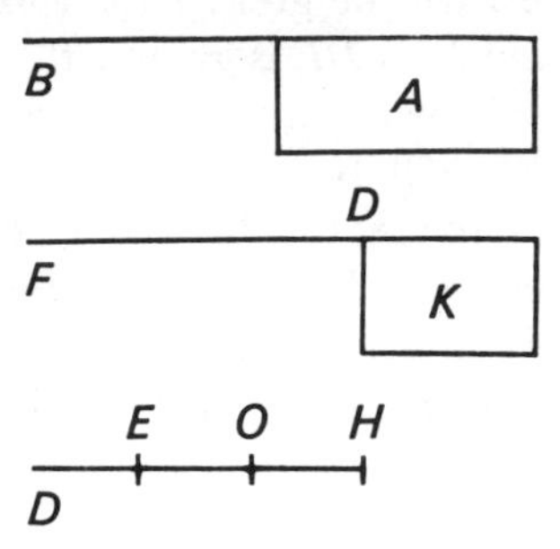

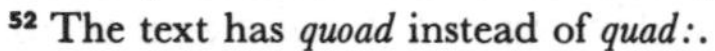
[52] The text has *quoad* instead of *quad:*.

CHAPTER II

On the Total Number of Rules

1. The rules which, generally speaking, it is convenient to know [for problems] up to and including solids extend to the cubic. The simple ones, since they are all of the same sort, we have contracted into one, even though this one might be extended to infinity. Those which deal with a constant, square, and first power are three in number, and although one of these may give rise to two solutions, since these are linked together, we count them as three. There are also three involving the cube, first power, and constant, but since one of these has two solutions, they count as four, and there are the same number for the cube, square, and constant. Up to this point, then, there are 12. The rules for the cube, square, first power, and constant are seven, for four of which there are twin solutions, so there are 11 in all. Therefore the primary and general cases total 23. Skipping the first of these, two derivatives may be joined to each of the others, one being in the nature of the square, the other of the cube. Thus there are 44 general derivatives. In addition, there are two [rules] involving a [second] unknown quantity, one when it is multiplied, the other when it is taken by itself.[1] Further, there is one general rule for the mean. The total of all of the primary cases worth knowing is 26 which, with 44 derivatives, gives us a complete collection of 70. Beyond these, there are a number of other particular [rules] which we add but, since the pleasure of knowing them is greater than the necessity, we will not number them with the others.

2. The importance of these can be summarized thus: When lines are derived from surfaces or surfaces from lines, [the rules for] the square, first power and constant are appropriate, and if [it is a matter of] the side of a tetragon or solid the case is one of the simple ones. If, however, two out of three [factors] are supposed to be unknown and

[1] 1663 omits "by itself."

they pertain to surfaces and lines, the cases of the first power and a [second] unknown quantity must be explored — in simple form if lines are being compared with lines, in multiplied form if lines are being compared with surfaces. But if bodies are being compared with lines, [the rules are those dealing with] the cube, first power, and number, [2]and if surfaces are being compared with bodies, it is those of the cube, square, and number.[2] If, moreover, it is the relation of surfaces, bodies, and lines that is sought, the rules for the cube, square, first power, and number are very useful. Furthermore, in all these cases, the comparison is always made with the number. It is also clear that, although it will sometimes be necessary to utilize all of these rules in a single instance, it will be worth the effort to take them up individually and to subjoin the derivatives to the primitive rules. Here they are:[3]

Primitive Cases Lacking Derivatives

1. $N = ax$
   $N = ax^2$
   $N = ax^3$
   $N = ax^4$
   $N = ax^5$[4]
   and so on, comparing the number with each power.

2.[5] $N + x^2 = ax$
   $N + x^3 = ax$
   $N + x^3 = ax^2$
   $N + x^4 = ax$
   $N + x^4 = ax^2$
   $N + x^4 = ax^3$
   $N + x^5 = ax$, or ax^2, or ax^3
   and so on without end.

3. N, x, and y

4. N, x^2, and y
   N, y^2, and x
   N, x^2, [and] y^2
   N, xy, and x or y or x^2 or y^2.

Primitive Cases	Derivative Cases
1. $N = x^2 + ax$	1. $N = x^4 + ax^2$ 2. $N = x^6 + ax^3$

[2] 1663 omits this material.

[3] The literal coefficients are, of course, not Cardano's. They have been added here and elsewhere throughout the book for the sake of clarity and, in many instances, to express that which Cardano frequently indicates by using the plural forms of his words for these terms.

[4] *nomini seu relato primo.*

[5] It is difficult to see why Cardano lists some of these cases with those "lacking derivatives"; *cf.* items 3, 6, 7, 10, and 11 in the following section.

Primitive Cases	Derivative Cases
2. $N + ax = x^2$	3. $N + ax^2 = x^4$ 4. $N + ax^3 = x^6$
3. $N + x^2 = ax$	5. $N + x^4 = ax^2$ 6. $N + x^6 = ax^3$
4. $N = x^3 + ax$	7. $N = ax^2 + x^6$ 8. $N = ax^3 + x^9$
5. $N + ax = x^3$	9. $N + ax^2 = x^6$ 10. $N + ax^3 = x^9$
6. $N + x^3 = ax$ (I)	11. $N + x^6 = ax^2$ (I) 12. $N + x^9 = ax^3$ (I)
7. $N + x^3 = ax$ (II)	13. $N + x^6 = ax^2$ (II) 14. $N + x^9 = ax^3$ (II)
8. $N = ax^2 + x^3$	15. $N = ax^4 + x^6$ 16. $N = ax^6 + x^9$
9. $N + ax^2 = x^3$	17. $N + ax^4 = x^6$ 18. $N + ax^6 = x^9$
10. $N + x^3 = ax^2$ (I)	19. $N + x^6 = ax^4$ (I) 20. $N + x^9 = ax^6$ (I)
11. $N + x^3 = ax^2$ (II)	21. $N + x^6 = ax^4$ (II) 22. $N + x^9 = ax^6$ (II)
12. $N = ax + bx^2 + x^3$	23. $N = ax^2 + bx^4 + x^6$ 24. $N = ax^3 + bx^6 + x^9$
13. $N + ax = bx^2 + x^3$	25. $N + ax^2 = bx^4 + x^6$ 26. $N + ax^3 = bx^6 + x^9$
14. $N + ax + bx^2 = x^3$	27. $N + ax^2 + bx^4 = x^6$ 28. $N + ax^3 + bx^6 = x^9$

Primitive Cases

15. $N + bx^2 = ax + x^3$ (I)

16. $N + bx^2 = ax + x^3$ (II)

17. $N + x^3 = ax + bx^2$ (I)

18. $N + x^3 = ax + bx^2$ (II)

19. $N + ax + x^3 = bx^2$ (I)

20. $N + ax + x^3 = bx^2$ (II)

21. $N\,[+]\,bx^2 + x^3 = ax$ (I)

22. $N + bx^2 + x^3 = ax$ (II)

Derivative Cases

29. $N + bx^4 = ax^2 + x^6$ (I)
30. $N + bx^6 = ax^3 + x^9$[6] (I)

31. $N + bx^4 = ax^2 + x^6$ (II)
32. $N + bx^6 = ax^3 + x^9$ (II)

33. $N + x^6 = ax^2$[7] $+ bx^4$ (I)
34. $N + x^9 = ax^3 + bx^6$ (I)

35. $N + x^6 = ax^2 + bx^4$ (II)
36. $N + x^9 = ax^3 + bx^6$ (II)

37. $N + ax^2 + x^6 = bx^4$ (I)
38. $N + ax^3 + x^9 = bx^6$ (I)

39. $N + ax^2 + x^6 = bx^4$ (II)
40. $N + ax^3 + x^9 = bx^6$ (II)

41. $N + bx^4 + x^6 = ax^2$ (I)
42. $N + bx^6 + x^9 = ax^3$ (I)

43. $N + bx^4 + x^6 = ax^2$ (II)
44. $N + bx^6 + x^9 = ax^3$ (II)

[6] 1570 and 1663 have x^3.

[7] 1570 and 1663 omit ax^2.

CHAPTER III

On Solutions in Simple Cases

1. The value of x is the quantity by which we prove the truth of those things which are proposed in a case or problem. For example, if someone says, Break 10 into two parts and square each of them, and let the difference between the squares be 60. Since we do not know which quantity is the greater and which the smaller, we let the smaller be the unknown, which we call x. Then the greater will be the remainder of 10, namely $10 - x$. We now follow through with what was proposed and multiply[1] each part by itself, making the square of the smaller x^2 and the square of the greater $x^2 + 100 - 20x$. We add the negative to the other part, making $x^2 + 100$ for one and $x^2 + 20x$ [for the other]. The difference between these is 60, by supposition. So we add 60 to the smaller, whence

$$x^2 + 100 = x^2 + 20x + 60.$$

We now subtract $x^2 + 60$ from both sides, leaving

$$20x = 40,$$

since if equals are subtracted from equals the remainders are equal. Dividing 40 by 20, the coefficient of x, gives us 2, the value of x. We now verify with this 2 the true nature of the problem that was posed. If the square of this, namely 4, is subtracted from 64, the square of 8, which is the difference between 10 and 2, there is left 60, the given number. What is proposed in the equation is also verified by 2, for its square, which is 4, added to 100 is equal to the square of the unknown, which is 4 again, plus $20x$, which is 40, plus 60. In either case they add up to 104. We say, therefore, with merit on account of these two [tests], that 2 is the value of x, and if you proceed correctly, both [types of] proof with a solution or equation will succeed.

[1] 1545 and 1570 have *ducemus*; 1663 has *dumus*.

Demonstration

2. So that the truth of this may be seen more clearly, and along with this its reason (for to know by demonstration is to understand, as has been said) let, for example,

A C D B

$$3x^3 = 24$$

and let AC be the side of one cube, CD of another, and DB of the third. Since, therefore, the cubes are equal to each other, the lines AC, CD, and DB will also be equal. Now divide 24, which is[2] the sum of the cubes, by the number which results from dividing AC into AB, which is 3, [and][3] it appears from V, 19 or VII, 17 of the *Elements* and XI, 31 of the same[4] that the cube of AC is 8. Therefore the side AC will be 2, the value of x. From this the following general rule may be gathered:

Rule

3. If there is no number, reduce [one of] the two given terms to a number by dividing both equally and, when one has become a number and the other a power, divide the number by the coefficient of the power. The result is the value of the power. If this power is now x, you have the value of x. If it is another power, take the root[5] of the number in accordance with the nature of the power: if it is a square, take the square [root]; if a cube, the cube root; if a fourth power, the root of the root; and so on, and this root is the true value of x.

For example,

$$20x^3 = 180x^5.$$

Since there is no number here, let the lower power (the cube) be a simple number, namely 20, and divide the greater or higher power (x^5) by the cube, making $180x^2$. Then divide the number 20 by 180, the coefficient of x^2. The value of x^2 comes out as $\frac{1}{9}$. But we are looking for the value of x, not of x^2. Take, therefore, the square root of $\frac{1}{9}$, which is $\frac{1}{3}$, as the true value.

Another example,

$$7x^2 = 21x^6.$$

[2] 1545 and 1570 have *et*; 1663 has *quod est.*

[3] 1545 and 1570 omit this word; 1663 includes it.

[4] The portions of Euclid to which Cardano refers here and elsewhere in *The Great Art* are collected in the Appendix to avoid repetition.

[5] *latus seu radicem.*

Reduce equally until [one term is] a number, making

$$7 = 21x^4.$$

Divide 7 by 21 and there arises $\frac{1}{3}$. Hence $\sqrt[4]{\frac{1}{3}}$, which is the side of the square of a square, is the value of x.

Still another,

$$2x^3 = 20x^4.$$

Having reduced the cube to a number,[6] the squares of the square become x's. Hence

$$20x = 2.$$

Divide 2 by 20. The result is $\frac{1}{10}$ and, since you have already divided by the coefficient of x, the value of x will be $\frac{1}{10}$.

Another,

$$20 = 5x^2.$$

Divide 20 by 5, resulting in 4 as the value of x^2. Hence the value of x is 2.

4. And in order to satisfy future cases, [the rule is:] Divide the coefficient of the greatest power — by the greatest, I mean the highest — into all the other terms and depress [all of them] by the lowest power. In this way, you will be following the rule of this chapter.

For example, let

$$4x^3 = 12x^2 + 8x.$$

The lowest power is x and the coefficient of the highest is 4. Divide through, therefore, by 4[x] and you have

$$x^2 = 3x + 2.$$

From all of this it is clear that a simple x[7] is far more appropriate than the negative. For powers extend to squares, cubes, and others, and their solutions will be roots. These are completely useless if x is negative. What has thus far been said of a number equal to the first power is generally true of any false x. As is shown in the first example, no false x can be found which the squares of the parts of 10, the difference between these squares being 60, will verify,[8] as was there proposed.

[6] 1663 omits this phrase.

[7] Better, a simple value (meaning the plus value) for x.

[8] 1545 has *variant*; 1570 has *veriant*; 1663 has *faciant*.

CHAPTER IIII[1]

On the General and Particular Solutions That Follow

1. Solutions by which no class of cases[2] can be completely resolved are called particular. Such are whole numbers or fractions or roots (square, cube, or what you will) of any number or, as I will put it, any simple quantity. Likewise, all *constantes* made up of two roots, either of which is a square root, a fourth root or, generally, any even root and, therefore, those consisting of two [such] terms and their *apotomes* (or, as they are called, *recisa*) of the third and sixth classes are not suitable for a general solution.[3]

[1] 1545 and 1570 use this old form of the Roman numeral; 1663 uses IV.

[2] A good deal of the meaning of this first paragraph turns on the proper translation of *capitulum*, which is elsewhere rendered as "case" or, sometimes, "rule," "proposition," or even "equation." Here, and perhaps elsewhere, it means a class of cases. (*Cf.* the definition of the Italian *capitolo* as given in Tommaseo e Bellini, *Dizionario della Lingua Italiana: Anticamente chiamavansi Capitoli di algebra i diversi Casi in cui si distingueva la risoluzione delle equazioni di secondi grado.*) Even so, Cardano cannot quite mean what he says in this paragraph, for it is obvious that some of the *capitula* listed in the preceding chapter are solvable by simple quantities. I take it, therefore, that he is here discussing the types of *capitula* to be encountered in the following chapters. The title of this Chapter IIII so indicates.

[3] The terms *constans* and *apotome* (frequently *binomium* and *recisum* elsewhere in *The Great Art*) are not easy to put into English. If "binomial" still meant only $a + b$ and not $a \pm b$, and if "residual" were in common use for $a - b$, these would be good terms to use. As it is, however, they are not.

In Chapter 3 of his *Ars Magnae Arithmeticae*, which is appended to *The Great Art* of 1663, Cardano outlines and illustrates the six types of *constans* and *recisum* to which he refers in this chapter. All consist either of an integer and square root or of two square roots. The first three are those in which the square of the greater component exceeds the square of the smaller by a number the square root of which is commensurable with the greater. The second three are those in which the square of the greater component exceeds the square of the smaller by a number the square root of which is incommensurable with the greater. He illustrates the six types thus:

	Constans (*binomium*)	*Apotome* (*recisum*)
1	$3 + \sqrt{5}$	$3 - \sqrt{5}$

2. Every case which consists of a number, square, cube, and first power has general solutions which can be derived from the [solution for the] case to which it reduces by adding or subtracting one-third the coefficient of the square, as will be shown in the proper place.

3. The general solutions of cases are —

for cases of the square equal to the first power and number, a two-term *constans* of the second class, as $\sqrt{19} + 3$;

for cases of the square and first power equal to the number, an *apotome* of the second class, as $\sqrt{19} - 3$;

for cases of the square and number equal to the first power, a two-term *apotome* and *constans* of the first class, as $3 + \sqrt{2}$ and $3 - \sqrt{2}$.

Where I speak of the first class, I understand also the fourth class, and where the second, the fifth, both with respect to two-term *apotomes* and with respect to two-term *constantes*.[4]

4. But a solution in the form of a single universal root [i.e., the root of a quantity], either square or cube, is suitable for the derivative cases, their principal cases being those for which the square or cube of the root serves as a solution. Thus, if the value of x in

$$x^2 = ax + N$$

is $\sqrt{19} + 3$, the solution in the case of

$$x^6 = ax^3 + N,$$

the coefficients and constants being the same,[5] will be $\sqrt[3]{\sqrt{19} + 3}$.

5. And as a square root can never be associated with anything except a number if it is to serve as a general solution so, contrariwise, a cube

---

2	$\sqrt{12} + 3$	$\sqrt{12} - 3$
3	$\sqrt{18} + \sqrt{10}$	$\sqrt{18} - \sqrt{10}$
4	$3 + \sqrt{2}$	$3 - \sqrt{2}$
5	$\sqrt{11} + 2$	$\sqrt{11} - 2$
6	$\sqrt{7} + \sqrt{3}$	$\sqrt{7} - \sqrt{3}$

He also points out that if the square of the greater component times the difference between the square of the greater and the square of the less is itself a perfect square, the *binomium* or *recisum* belongs to one of the first three types, otherwise to one of the second three.

[4] 1545 has *tam in apotome quam ex duobus nominibus constante*; 1570 and 1663 have *tam in apotome quam in ea quae ex duobus nominibus constant.*

[5] 1545 has *sub eadem quantitate existentibus*; 1570 and 1663 omit *existentibus.*

root must be joined to a cube root and not to a number to serve as a general solution. When, for instance, [cube root] is joined [to cube root] it produces a solution (not a whole number, however) for

$$x^3 = ax + N$$

and one of these terms subtracted from the other will be a solution for the case of

$$x^3 + ax = N.$$

Thus, if $\sqrt[3]{4} + \sqrt[3]{2}$ is a solution for the case of

$$x^3 = ax + N,$$

$\sqrt[3]{4} - \sqrt[3]{2}$ is a solution for the case of

$$x^3 + ax = N.$$

6. The case of

$$x^3 = ax^2 + N$$

has as its solution a *constans* of three terms in continued proportion,[6] the two extremes of which are cube roots and the middle one a number, like $\sqrt[3]{16} + 2 + \sqrt[3]{4}$. And the case of

$$x^3 + ax^2 = N$$

has a solution similar in all respects to the preceding except that the [whole] number is negative, like $\sqrt[3]{16} - 2 + \sqrt[3]{4}$.

7. This must also be understood: [If] simple roots are to be had for general solutions, the cases must also be simple. So a cube root will serve in the case of a number equal to a cube, and a square root in the case of a number equal to the second power, and the 5th root in the case of a number equal to the 5th power.[7] And just as these simple [roots] will not serve for compound cases, so no solution compounded from several incommensurable roots[8] can serve in the simple case.

[6] 1545 has *habet aequationem constantem in tribus quantitatibus proportionalibus*; 1570 and 1663 have *habet aequationem quae constat ex tribus quantitatibus in continua proportione.*

[7] *et relata, capitulo relati aequalis numero.*

[8] 1545 has *ex pluribus radicibus incommensurabilibus*; 1570 and 1663 have *ex pluribus radicibus incommensi.*

CHAPTER V

Showing the Solution of Cases Composed of Minors, Which Are the Square, Constant, and First Power

Demonstration

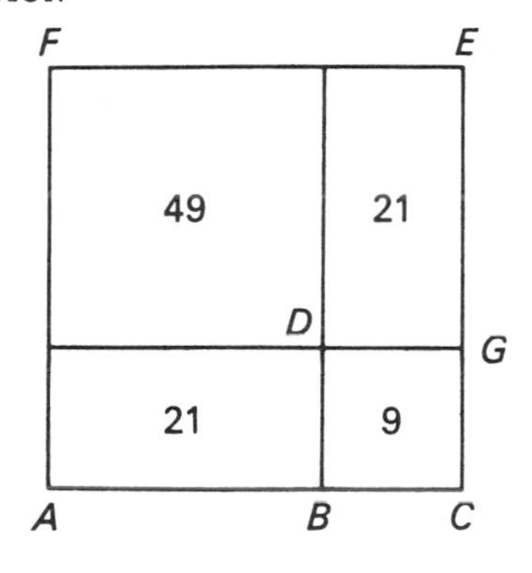

Let the square *FD* and $6x$, for example, equal 91. Then I produce *DB* and *DG*, which are 3, one-half of 6, the coefficient of x,[1] and complete the square *DGBC* and then, having produced *CG* and *CB*, I perfect[2] the square *AFEC*, just as in II, 4 of the *Elements*. Since, therefore, *DB* times *AB*, from the definition in the second book of the *Elements*, gives the surface[3] *AD* and since the product of any coefficient and the value of x gives the value of that number of x's (as, if x is 4 and there are 5 x's, the 5 x's will be 20 and so much will therefore be produced by 4, the value of x, times 5, the coefficient of x, as we have shown in the third chapter), therefore when *BD* is 3 and *AB* is the value of x, the surface *AD* will be equal to $3x$ or the value of $3x$. But the surface *DE* is equal to *AD*, according to I, 43 of the *Elements* and it, therefore, is the value of another 3 x's. Hence the two surfaces, *AD* and *DE*, are equal to $6x$, wherefore these plus the square *FD* are 91. But the square *CD* is 9, since *BD* is 3. Therefore the square of *AC* is 100 and its side, *AC*, is 10. Since, therefore, *BC* is 3, subtracting *BC* from *AC* leaves *AB*, the side of *DF*,[4] as 7.

[1] 1545 has *tunc faciam DB et DG cum fuerint productae esse 3, dimidium scilicet 6, numeri rerum*; 1570 and 1663 have *tunc productam db et dg quae sint 3, dimidium 6, numeri rerum.*

[2] 1570 and 1663 have *perficiam;* 1545 omits any verb.

[3] 1570 and 1663 omit this word; in 1545 it appears as *super finem*, but this is corrected by an erratum note at the beginning of the book.

[4] 1570 and 1663 have *DE.*

ANOTHER DEMONSTRATION

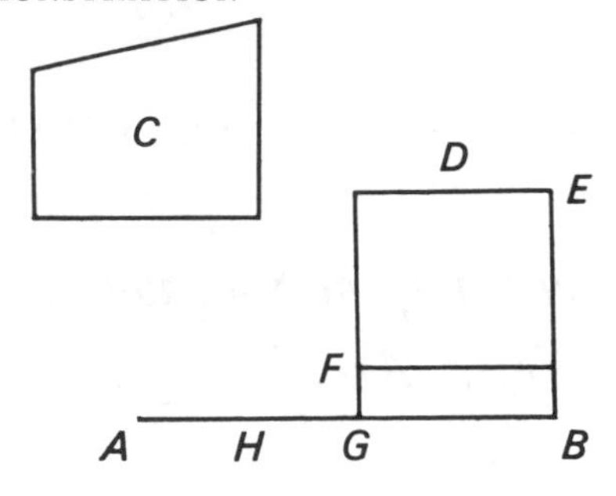

2. In this instance, let AB be the coefficient of the x's which are equal to C, the constant, and D, the second power. I square BG — BG being half of AB — thus making GE, and from it I subtract C, the constant, the surface EF being equal to the constant C, and I mark off the square root of the surface FB, which is GH. I say that either line, BH or HA, is the root of the square D,[5] whence it follows that there are two true solutions for this case and that their[6] sum is equal to the coefficient of x, namely AB, for it is true that the rectangle $AH \times HB$ plus the square of GH is equal to the square of BG, according to V, 2 of the *Elements*. The square of HG, moreover, is equal to the surface FB. Therefore the rectangle $AH \times HB$ is equal to EF and therefore to C, the constant. The product of AB and HB, however, according to II, 3 of the *Elements*, is equal to the square of HB and the rectangle $AH \times HB$. Therefore the product of AB, the coefficient of x, and HB, the value of x, is equal to the number C and the square of HB, as was to be proved. And, by similar reasoning, the rectangle $AB \times AH$ is equal to the square of AH and the product of AH and HB. But $AH \times HB$, as has been proved, is C, the constant. Therefore the rectangle $AB \times AH$ — that is, the product of the coefficient of x and the value of x — is equal to the square of x and the given number.

From this it is clear that they are mistaken who say that if BH, for instance, is the value of x and GF is 3, the rectangle $BH \times GF$ will be $3BH$[7] or triple BH.[7] For it is impossible that a surface should be composed of lines, whatever their number or other proportion, since there may be infinite lines on a surface, inasmuch as any continuous quantity admits of no end to its division. But the truth is that if GF were comprised of three units,[8] for example — that is, three parts — BH would have as many parts as there are units in the number which

[5] 1545 has *dico lineas BH et HA esse utrasque latera quadrati D*; 1570 and 1663 have *dico utranque lineam bh et ha esse latus quadrati d.*

[6] 1545 and 1570 have *quarum*; 1663 has *quare.*

[7] The text has GH.

[8] 1545 has *unitates;* 1570 and 1663 have *monades.*

it is said to contain.[9] Thus if we say that *BH* is 12, *GF* will be 3, when *GF* is a fourth of *BH*, and thus it is true that the product of *BH* and *GF* is a surface containing 36 square surfaces, each of which has a tetragonic side of unity — that is, one of those parts which arise when *BH* is divided into 12 and *GF* into 3. This is shown beautifully by Plato in the *Meno* with respect both to rational numbers and to surds.[10]

Now do not be astonished at this second demonstration, which is explained differently than by Mahomet, for [our] unchanged figure shows more about the matter [than his] though perhaps more obscurely except in one respect among many. We employ it both because of its convenience and brevity and because, by it, we are able to show both values in one demonstration.

Another Demonstration[11]

3. Let the square *AC* in the third figure equal $6x$ plus 16, and assume that *AD* is the coefficient of x, namely 6. Then the surface *AH* is $6x$, wherefore *DC*, the remainder, will be exactly 16. Let *AD* be divided equally at *G* and construct the squares of *GB* and *GD*, which are *GK* and *GE*. Since, therefore, *BC* is equal to *BA* and *BK* to *BG*, *KC* will be equal to *GA* and therefore, also, to *GD* and *FL*. Since *DE* and *DG* are equal to each other

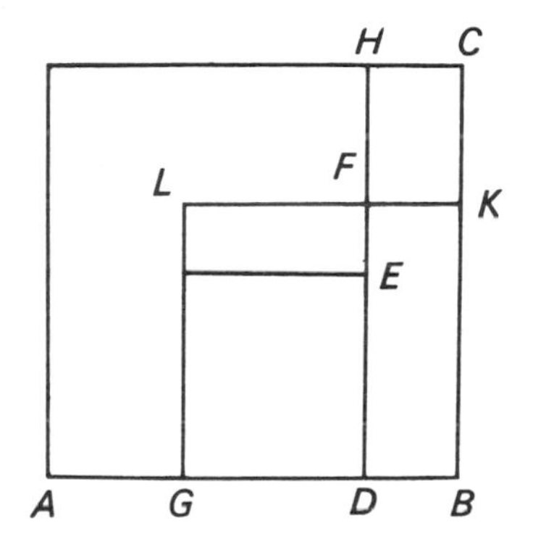

and *DF* and *BG* are likewise equal, *FE* will be equal to *DB* and, therefore, also to *FK*. Hence the two lines *FK* and *FH* are equal to *FL* and *FE* and the angles at *A*, *D*, and *F* are right angles. Therefore the surface *FC* is equal to *LE*. But *FC* plus *FB* equals 16 and, therefore, *LE* plus *FB* equals 16 and, by adding the square *GE*, which is 9

[9] The sense of this sentence is clear enough, but the grammar of the original is quite obscure: *sed veritas est, quod si GF contineat tres unitates (gratia exempli) id est partes tres lineae BH, divisae in tot partes, quot unitates sunt in numero quem dicitur continere.*

[10] 1545 has *hoc autem tam in rationalibus, quam in irrationalibus pulchre ostendit Plato in Memnone*; 1570 and 1663 substitute *tam in rhetis quam in alogis* for *tam in rationalibus quam in irrationalibus*. *Quaere* the correctness of *Meno* as a translation for *Memnone* or, if it is correct, of Cardano's reference to it. But what else could it be or to what else might he be referring? Could *in Memnone* be a misprint for, say, *in mnemone* or something of the sort and thus be translatable as "if I remember correctly"?

[11] The lettering of the bottom of the diagram is reversed in 1570 and 1663 from that given in 1545 and used here.

(since *GD* was 3), the square *GK* will be 25. Hence the side, *GB*, will be 5 and this, plus *GA*, which is 3, makes *AB* a total of 8, the value of x.

4. In accordance with these demonstrations, we will formulate three rules and we attach this jingle in order to help remember them:

Squeaxno, adtwix
Noesquax, adsub
Axesquno, subadsub.[12]

RULE I

There is one thing common to all these cases — that one-half the coefficient of x is multiplied by itself. When, therefore, the square is equal to the first power and constant, this is signified by Squeaxno. You can understand this much from the first word [of the jingle] or you can figure it out by following the sounds beginning with the first letter. Thus

Squeaxno signifies $x^2 = ax + N$
Noesquax signifies $N = x^2 + ax$
Axesquno signifies $ax = x^2 + N$

In Squeaxno, therefore, or the case of the square equal to the first power and number, add the square of one-half the coefficient of the first power to the constant of the equation and take the square root of the whole. To this add one-half the coefficient of the first power, and the sum is the value of x.[13]

For example, let

$$x^2 = 10x + 144.$$

Multiply 5 by itself, making 25, the square of one-half the coefficient of the first power. To this add 144, making 169, the square root of which is 13. Add 5, one-half the coefficient of x, to this, making 18 the value of x.

[12] In the Latin these are:

Querna, da bis
Nuquer, admi
Requan, minue dami

Cf. the comment of M. M. Marie on this jingle in his *Histoire des Sciences Mathématiques et Physiques* (Paris, 1883), p. 253: "C'est parfait! pourvu que l'élève ne *requane* pas quand il s'agirait de *querner* ou de *nuquer*."

[13] I.e., if $x^2 = ax + N$, $x = \sqrt{N + (a/2)^2} + a/2$.

Again, let

$$x^2 = \tfrac{2}{3}x + 11.$$

Multiply $\frac{1}{3}$, which is half the coefficient of x, by itself, making $\frac{1}{9}$. Add 11 to this, making $11\frac{1}{9}$. Take the square root, which is $3\frac{1}{3}$, and add to this $\frac{1}{3}$, one-half the coefficient of x, making $3\frac{2}{3}$, the value of x.

Again, let

$$x^2 = 10x + 6.$$

Multiply 5, one-half the coefficient of x, by itself, making 25. Add 6 to this, making 31. Add 5, half the coefficient of x, to the square root of this, and the value of x will be $\sqrt{31} + 5$.

Again, let

$$x^2 = \sqrt{12}x + 22.$$

Multiply $\sqrt{3}$ by itself, making 3, the square of one-half the coefficient of x. Add 22 to it, making 25, the square root of which is 5. To this add $\sqrt{3}$, which is half the coefficient of x, making the value of x $5 + \sqrt{3}$.[14] If, in this case, the constant had been 20, the value of x would have been $\sqrt{23} + \sqrt{3}$,[15] and if it had been 9, the value would have been $\sqrt{12} + \sqrt{3}$, which is to say $\sqrt{27}$.

And if

$$x^2 = \sqrt{12}x + \sqrt[3]{10},$$

multiply, as before, $\sqrt{3}$, one-half the coefficient of x, by itself, making 3; add to this $\sqrt[3]{10}$, making $3 + \sqrt[3]{10}$. Take the root of this, which is $\sqrt{3 + \sqrt[3]{10}}$, and add to it half the coefficient of x, making the value of x $\sqrt{3} + \sqrt{3 + \sqrt[3]{10}}$.

We have used this variety of examples so that you may understand that the same can be done in other cases and will be able to try them out for the two rules which follow, even though we will there be content with only two examples. It is clear, therefore, that we here add twice, namely the number to the square of one-half the coefficient of x and one-half the coefficient of x to the root of the [first] sum, and this is what we said in the jingle — adtwix, or add twice.

Rule II

5. If, however, the number is equal to the square and first power, add the number to the square of one-half the coefficient of x, take the

[14] 1570 and 1663 have $5 + \sqrt{5}$.

[15] 1663 has $\sqrt{23} + 3$.

square root of the whole, subtract from it one-half the coefficient of x, and the remainder is the value of x.[16]

For example,

$$144 = 10x + x^2.$$

Multiply 5, one-half of 10, the coefficient of x, by itself, making 25. Add 144 to this, making 169, the square root of which is 13. Subtract 5, one-half the coefficient of x, from this, leaving 8 as the value of x.

Again, let

$$6 = 10x + x^2.$$

Multiplying 5, one-half the coefficient of x, by itself makes 25. Add 6, making 31. From the square root of this subtract 5, one-half the coefficient of x, making $\sqrt{31} - 5$ the solution.

Corollary. From this it appears that this rule differs from the preceding only in that one-half the coefficient of x is subtracted from the root of the sum instead of being added to it. And this is what we said in the jingle: adsub, that is add first, then subract — namely, add the number to the square and then subtract one-half the coefficient of x from the root of the sum.

Corollary. From this it is apparent that the difference between the solution for the square equal to the first power and number and for the number equal to the first power and square is precisely the coefficient of x, where the number and the coefficient are the same in the two cases. Thus, in

$$x^2 = 10x + 144,$$
$$x = 18$$

and in

$$144 = x^2 + 10x$$
$$x = 8$$

and the difference between 18 and 8 is 10.

Rule III

6.[17] If the first power is equal to the square and number, multiply as before one-half the coefficient of the first power by itself and, having subtracted the number from the product, subtract the root

[16] I.e., if $N = x^2 + ax$, $x = \sqrt{N + (a/2)^2} - a/2$.

[17] 1663 omits this section number.

of the remainder from one-half the coefficient of the first power or add the two of them, and the value of x will be both the sum and the difference.[18]

For example,

$$x^2 + 16 = 10x.$$

Multiply 5 by itself, making 25, as before. Then subtract 16 from 25, leaving 9, the square root of which is 3. Having added this to or subtracted it from 5, one-half the coefficient of x, the values of x turn out to be 8, when added, and 2, when subtracted. If, therefore, 10 x's, each of which is 2, are added together, they will be 20, and this will also be 16 plus the square of 2. Again, if 10 x's, each of which is 8, are added together, they will be 80, and so much also is the square of 8 when 16 is added to it.

Again, if I say

$$10x = x^2 + 6,$$

multiplying 5, one-half the coefficient of x, by itself makes 25 and subtracting 6 from this leaves 19, the square root of which added to or subtracted from 5, gives us the values of x, the greater being $5 + \sqrt{19}$[19] and the smaller $5 - \sqrt{19}$.

Note. If the number cannot be subtracted from the square of one-half the coefficient of the first power, the problem is itself a false one and that which has been proposed cannot be. It must always be observed as a general rule[20] throughout this treatise that, when those things which have been directed cannot be carried out, that which is proposed is not and cannot be. Now, however, we subjoin several problems, two from Mahomet, the remainder our own. They are the most difficult of all those which do not use a multiple x or a special[21] rule.

Problem I[22]

There is a number from the square of which, if you subtract one-third and one-fourth of the same square and, in addition, 4, and

[18] I.e., if $x^2 + N = ax$, $x = a/2 \pm \sqrt{(a/2)^2 - N}$.

[19] 1570 and 1663 have $5 + \sqrt{10}$.

[20] 1545 has *pro regula universali*; 1570 and 1663 have *pro regula generali*.

[21] 1545 has *particulari*; 1570 and 1663 have *propria*.

[22] Of the various available versions of al-Khowarizmi cited above, p. 8, this problem appears only in that of Libri at pp. 293–294. A very similar problem is posed by Fibonacci (cited p. 8 above), I, 422–423.

multiply the remainder by itself, arises a product equal to the square of the same number plus 12. Let the square of the unknown number which you seek be x. Subtract one-third and one-quarter of the same[23] and 4 more. This makes $\frac{5}{12}x - 4$. Multiply this by itself, making $\frac{25}{144}x^2 + 16 - 3\frac{1}{3}x$, and this is equal to $x + 12$. Subtracting similars,

$$x = \frac{25}{144}x^2 + 4 - 3\tfrac{1}{3}x.$$

Move the negative quantity to the other side, in accordance with the universal rule, and

$$4\tfrac{1}{3}x = \frac{25}{144}x^2 + 4.$$

Then, in accordance with the fourth rule of Chapter III, divide[24] the coefficient of the first power and 4 by $\frac{25}{144}$, the coefficient of x^2. Thus

$$24\tfrac{24}{25}x = 23\tfrac{1}{25} + x^2,$$

whence, in accordance with the third rule [of this chapter], multiply $12\frac{12}{25}$ by itself, making $155\frac{469}{625}$; subtract $23\frac{1}{25}$, making $132\frac{444}{625}$,[25] the square root of which is $11\frac{13}{25}$; add to this $12\frac{12}{25}$, one-half the coefficient of x, giving 24 as the value of x. Hence the square root of this is the number which was sought. From this we learn how to avoid the derivative cases by [using] the principal ones, for if, instead of x for the first number, we had used x^2 as the fundamental [term] of the operation, we would have arrived at

$$x^4 + 23\tfrac{1}{25} = 24\tfrac{24}{25}x^2.$$

So let this be an example to you. Now follows his [Mahomet's] second [problem].

Problem II[26]

There were two leaders each of whom divided 48 *aurei* among his soldiers.[27] One of these had two more soldiers than the other. The one who had two soldiers fewer had 4 *aurei* more [than the other] for

[23] 1663 omits this word.

[24] 1545 has *devidi*; 1570 and 1663 have *divisi.*

[25] The text has $132\frac{544}{625}$.

[26] This precise problem occurs in none of the versions of al-Khowarizmi cited above, page 8. One entirely similar to it in substance and principle, however, is given on p. 286 of Libri and pp. 119 and 157 of Karpinski. Similar problems are also posed by Fibonacci, I, 413 ff.

[27] 1545 has *Duo duces diviserunt militibus suis aureos 48 singuli*; 1570 and 1663 have *Fuerunt duo duces quorum unusquisque divisit militibus suis, aureos 48.*

each soldier. What is to be found is how many soldiers each had. Let the smaller number of soldiers be x. The greater will be $x + 2$. Since the sums to be distributed are the same, it is clear that the quantities will be in the same proportion. Four, however, is one-twelfth of 48. Therefore multiply $\frac{1}{12}$ by $x + 2$, making $\frac{1}{12}x + \frac{1}{6}$. Multiply this by the first number of men, making $\frac{1}{12}x^2 + \frac{1}{6}x$. Raise this whole sum to x^2 [i.e., multiply through by 12], making

$$x^2 + 2x = 24.$$

Take one-half the coefficient of x, which is 1, and the square of this is 1. Add 24, making 25, and from the square root of this subtract 1, one-half the coefficient of x, making 4, the smaller number of men, and 6, the greater. So the first has 12 *aurei* apiece, the other 8 *aurei*. The multiplication, however, by which the fraction of the square becomes an integer is made by 2 more than [the number of] men, namely by 12. And the reason for this is that the ratio of the second difference to the first is the ratio of the sum that is to be divided to the product of the [two] numbers of men. That is, the ratio of 48 to 24 (the product of 4 and 6) is that of 4 (the difference in the *aurei*) to 2 (the difference in the men). By this is shown the *modus operandi* in questions of proportions, particularly[28] when we wish a whole number, as in the number of men. In such cases it would be extremely absurd to arrive at half a man, let alone some irrational quantity or root.[29]

Problem III

Now let us pose our own problems, the first of which is similar to the preceding. There were two associations, one of which had three more members than the other. They divided equal numbers of *aurei* among their members. The *aurei* were 93 more than the total of the members in the two associations,[30] and the members of the smaller association each received 6 *aurei* more than the members of the larger association. Let the membership of the first association be x and that in the second will be $x + 3$. Therefore the sum of the *aurei*, which is 93 more than the membership in the two associations, is 96[31] $+ 2x$.

[28] 1545 has *sed magis praecipue*; 1570 and 1663 have *et praecipue*.

[29] 1545 has *nedum quantitatem aliquam irrationalem vel radicem*; 1570 and 1663 have *nedum quantitatem aliquam alogam seu latus*.

[30] 1545 has *93 plus aggregato hominum, in ambabus societatibus existentium*; 1570 and 1663 have *93 plus numero hominum ipsorum utriusque societatis simul iunctorum*.

[31] 1663 has 69.

The ratio of the excess of *aurei*, however, which the members of the smaller association received to the excess of men — i.e., 6 to $3x$ — is as the sum of the *aurei* to the product of the number of men in the first association and the number of men in the second. The ratio of 6 to 3, however, is duple. Therefore the ratio of $2x + 96$[32] to $x^2 + 3$ — the product of x and $x + 3$ — is duple. Hence one-half of $2x + 96$, which is x[33] $+ 48$, is equal to $x^2 + 3x$. Then, subtracting x from both sides yields

$$x^2 + 2x = 48.$$

Multiplying one-half of 2 by itself yields 1, for half of 2 is 1. Add 48 to this, making 49, the root of which is 7. From this subtract 1, one-half the coefficient of x, and you will have the value of x and the membership of the first association, namely 6. Hence the number of men in the second association is 3 more. If these two are added together and 93 more is added, the number of *aurei* is 108, giving 18 *aurei* for each member of the first organization and 12 for each member of the second.

Otherwise and more easily: In the operations shown, let x be as before. The sum of the *aurei* will be $2x + 96$. Divide by x and by $x + 3$, and you will have

$$\frac{2x + 96}{x} = 6 + \frac{2x + 96}{x + 3}$$ [34]

Subtracting $\frac{2x + 96}{x + 3}$ from $\frac{2x + 96}{x}$ leaves 6. But the same subtraction also yields $\frac{6x + 288}{x^2 + 3x}$ and this, therefore, is equal to 6. Dividing $6x + 288$ by 6 gives $x^2 + 3x$, for if I divide 10 by 2, 5 arises, and if I divide 10 by 5, 2 arises. Therefore dividing $6x + 288$ by 6 yields $x + 48$, and this is equal to $x^2 + 3x$, wherefore, as before, x is 6.

Problem IIII

There is a number which, if twice its square root is added to it and if to this sum twice its square root is added, gives a total of 10. Now you say that 10 is equal to the second number plus twice its square root. Let, therefore, the second number, the sum, be x^2 and this plus twice its square root is 10. Therefore the value of x, according to the

[32] 1570 and 1663 have 69.

[33] 1570 and 1663 have $\sqrt{x}$; 1545 uses *v: positio*, the *v:* presumably being intended for *una*.

[34] 1663 has 66.

second rule, is $\sqrt{11} - 1$. Hence subtracting twice this from 10 leaves the sum as $12 - \sqrt{44}$. This, however, by supposition, consists of a square and two roots. Hence

$$x^2 + 2x = 12 - \sqrt{44}.$$

Multiplying 1, one-half the coefficient of x, by itself makes 1. Add this to the number, making $13 - \sqrt{44}$. Take the square root of this and from it subtract 1, one-half the coefficient of x. You will then have $\sqrt{13 - \sqrt{44}} - 1$. Twice this amount subtracted from the sum leaves the number first proposed as $14 - \sqrt{44} - \sqrt{52 - \sqrt{704}}$, and thus you may proceed by regression as far as you wish from an original starting point.[35]

I will be more prolix here in the examples — these cases being very useful to the merchant — partly because they [the examples] serve to introduce tyros to these matters (and teachers ought always to teach their poor pupils a certain number of minutiae very thoroughly) and partly because we can quickly work out such minutiae through the remaining examples.

Problem V

Find the number which, when its cube root is subtracted from it and when the square root of the remainder is added to the remainder, yields the first number. Let the remainder after you have subtracted the cube root be x^2. Then add to this its square root, making $x^2 + x$, and this is equal to x^3, for the sum of x^2 and what was added is as much as there was at first. Hence what is added is equal to what was subtracted. The cube root of the whole quantity, however, is to be subtracted. Hence x is the cube root of the sum and therefore the sum is a cube and is equal to $x^2 + x$.[36] Divide through by x[36] and

$$x^2 = x + 1.$$

Hence x equals $\sqrt{1\frac{1}{4}} + \frac{1}{2}$. But the first number was x^3. Hence the first number is $\sqrt{5} + 2$.

Problem VI

A certain man went to market three times. The first time he brought back twice what he had taken with him. On the second trip he took

[35] *ab ultimo semper inchoando termino.*

[36] 1545 has the abbreviation *co:*, a misprint for *po:*, i.e., *positione*. In other contexts, *co:* could be an abbreviation for *cosa*, another term for x or the first power.

with him this double amount and returned with the same plus its square root plus 2 additional *aurei*. All this he preserved, and he returned to market with it a third time and his profit from this trip was the square of what he took with him and 4 *aurei* in addition. He returned, moreover, with 310 *aurei*. I want to know, therefore, how much he took with him on his first trip.

Say that he returned with 310 *aurei* and that this is equal to what he brought back from the second trip plus the square of this amount plus 4. Therefore, subtracting 4 from both sides, what he took with him on the third trip [plus] its square is equal to 306 *aurei*. We let x be the money he took with him. Thus we have

$$x^2 + x = 306.$$

Hence, according to the second rule, x equals $\sqrt{306\frac{1}{4}} - \frac{1}{2}$, which is 17, and this is the number of *aurei* he took with him on his third trip and the number had and saved from his second trip. It was said, however, that his profit from the second trip was the square root of what he took with him and 2 more. He returned with 17. Therefore, if his profit had been the square root alone, he would have returned with 15. So, assuming that x^2 is the amount of money which he took with him, we have

$$x^2 + x = 15.$$

Hence, by the second rule, x is $\sqrt{15\frac{1}{4}} - \frac{1}{2}$, and this plus 2 *aurei* was his profit on the second trip. His total profits on this trip, therefore, were $\sqrt{15\frac{1}{4}} + 1\frac{1}{2}$. He returned home, however, with 17 *aurei*. Hence he went with $15\frac{1}{2} - \sqrt{15\frac{1}{4}}$ and this is what he saved from his first trip and is twice what he went with on that trip. The first time he went to market, therefore, he took with him half of $15\frac{1}{2} - \sqrt{15\frac{1}{4}}$ *aurei*, which is $7\frac{3}{4} - \sqrt{3\frac{3}{16}}$ *aurei*.

Problem VII

A certain king sent 128,000 *aurei* to the proconsul who was leading his army so that he might hire 7000 foot soldiers and 7000 horsemen. The ratio of the stipend was such that 100 *aurei* would hire 18 more foot soldiers than it would mounted men. A certain tribune of soldiers came to the proconsul with 1700 foot men and 200 horsemen and asked for his share of the pay. This is similar to the third question.

Consider that 128,000 are 1280 hundreds. Since it was said that for each hundred *aurei* the difference between the number of foot soldiers and the number of horsemen is 18, divide 1280 into two parts, one of which multiplied by one quantity produces 7000 and the other multiplied by the same quantity plus 18 also produces 7000. Taking x, then, as the number of horsemen, the number of foot soldiers will be $x + 18$. Divide 7000 by each of these and their sum will be 1280, for if each of the parts of 1280 multiplied by x and by $x + 18$ makes 7000, then if 7000 is divided by x and by $x + 18$, the sum of the quotients will be 1280. From these two divisions come $\frac{14{,}000x + 126{,}000}{x^2 + 18x}$ and this is equal to 1280. Therefore, divide the numerator by 1280 and remove $x^2 + 18x$ [to the other side of the equation]. The result is

$$10\tfrac{15}{16}x + 98\tfrac{7}{16} = x^2 + 18x.$$ [37]

Therefore

$$x^2 + 7\tfrac{1}{16}x = 98\tfrac{7}{16}$$

and

$$x = \sqrt{110\tfrac{929}{1024}} - 3\tfrac{17}{32}.$$

But

$$\sqrt{110\tfrac{929}{1024}} = 10\tfrac{17}{32}.$$

So, subtracting $3\frac{17}{32}$, the value of x is 7, and so many horsemen will 100 *aurei* hire. They will likewise hire 25 foot soldiers. Therefore the pay for 1700 foot soldiers ought to be 6800 *aurei* and for 200 mounted men $2857\frac{1}{7}$ *aurei*.

Problem VIII

Divide 20 into three proportional quantities,[38] the second of which is equal to the sum of the roots of the first and third. Let the second be x; the remainder will be $20 - x$. Since two parts between which x will fall as a mean proportional are to be made from this, the product of the two will be x^2, in accordance with VI, 16 of the *Elements* and therefore, according to II, 5 of the *Elements* or the rules of the sixth

[37] The translation of the last few sentences is a little forced but correctly represents, I believe, the intent of the original: . . . *exeuntia iuncta faciunt 1280, ex* [1663 has *et*] *talium igitur divisione aggregantur* $\frac{\text{Pos. } 14000 \text{ p: } 126000}{1 \text{ quad. p: } 18 \text{ pos.}}$ *et hoc cum* [*sic*] *sit aequale 1280, igitur diviso numeratore per 1280, exit 1 qd*^{m} *p: 18 pos. facta igitur tali divisione, prodit* $10\frac{15}{16}$ *positionibus p:* $98\frac{7}{16}$, *hocque est aequale 1 quadrato p: 18 positionibus.*

[38] 1545 has *quantitates proportionales*; 1570 and 1663 have *quantitates analogas.* This difference in the language of the editions occurs frequently throughout the text and will, for the most part, not be further noted.

book,[39] we multiply one-half of $20 - x$ by itself, making $100 - 10x + \frac{1}{4}x^2$, from which we subtract x^2, making $100 - 10x - \frac{3}{4}x^2$. Add the square root of this to, and subtract it from, one-half of $20 - x$,[40] and you will have these parts: $10 - \frac{1}{2}x + \sqrt{100 - 10x - \frac{3}{4}x^2}$ and $10 - \frac{1}{2}x - \sqrt{100 - 10x - \frac{3}{4}x^2}$. In order to add the universal roots of these, do as I taught you in the third book: first add the two quantities and you will have $20 - x$, then multiply them by each other and add twice [the square root of] the product to their sum, making a total of $20 + x$, the square root of which is equal to x. Hence

$$x^2 = 20 + x.$$

Therefore, in accordance with the first rule, we multiply $\frac{1}{2}$, which is one-half the coefficient of x, by itself, making $\frac{1}{4}$; add to this 20, making $20\frac{1}{4}$; take the square root of this, which is $4\frac{1}{2}$; and add to it $\frac{1}{2}$, which is one-half the coefficient of x, making 5, the value of x, namely the mean quantity. Hence we divide the remainder of 20 into two parts between which 5 falls [as a mean proportional] and these will be discovered through another unknown, or by the rules of the sixth book, to be $7\frac{1}{2} + \sqrt{31\frac{1}{4}}$ and $7\frac{1}{2} - \sqrt{31\frac{1}{4}}$, the sum of the roots of which, when added together, is 5.

Problem IX

Divide 10 into two parts the greater of which minus twice its square root is equal to the smaller plus twice its square root.

It is clear that the difference between the greater and the smaller is twice the root of the greater plus twice the root of the smaller. Let this difference be $\sqrt{4x}$ and let one part be $5 + \sqrt{x}$ and the other $5 - \sqrt{x}$. Take the sum of the roots of these parts and this is, from the fourth book, the most universal root[41] $\sqrt{10 + \sqrt{100 - 4x}}$, and twice this is equal to $\sqrt{4x}$. Half of one is equal to half the other, namely $\sqrt{x}$ to the ultimate root. And the square of one is equal to the square of the other, namely,

$$x = 10 + \sqrt{100 - 4x}.$$

[39] 1545 has *quia igitur ex hoc facere oportet partes duas, inter quas positio cadat proportionalis, erit ex 16ª 6ⁱ elementorum, ut ex una in aliam fiat quadratum positionis, quare per 5ᵃᵐ 2ⁱ elementorum, seu ex regulis sexti libri*; 1570 and 1663 have *quia igitur ex hoc facere oportet partes duas, inter quas positio cadat proportione media, eritque ut ex una in aliam fiat quadratum positionis, quare per 16 6ⁱ Elementorum Ex 5ⁱ 2 Elementorum vel Reg: 6ⁱ libri.*

[40] The text has $20 + x$.

[41] 1545 has *radix universalissima*; 1570 and 1663 have *radix tota* (*quam universalissimam appellare solent*).

Hence

$$x - 10 = \sqrt{100 - 4x}.$$

The square of one is equal to the square of the other:

$$x^2 + 100 - 20x = 100 - 4x.$$

Hence x^2 is equal to $16x$ and x is equal to 16. But we wish the difference between the parts to be $\sqrt{4x}$. Hence the difference is $\sqrt{64}$, which is 8. Thus you have avoided the fourth power by using $\sqrt{x}$.

PROBLEM X

There were men in three organizations, and their numbers were in such proportion that the product of the numbers in the second and third was the total of those in all three plus the cube of the number in the first. You ought in this case to consider that it would be nonsense for the numbers to be surds[42] or fractions, for it is not proper to assume part of a man. See, therefore, in what proportion the square of half the product of the second and third is greater than the sum of all by some square number, and you will discover that in the duplicate — 1, 2, 4 — the square of one-half of 8, which is 4, [since 8] comes from 2 times 4, exceeds 7, the sum, by 9, a square number. This is what, beginning with some simple unknown, you will look for. Assume, then, a like number of x's for the numbers required — namely $1x$, $2x$, and $4x$, the sum of which is $7x$. Add x^3 to this, making $x^3 + 7x$, and this is equal to $8x^2$, the product of the second and third [of the series]. Divide everything by x, making

$$x^2 + 7 = 8x.$$

Hence, in accordance with the third rule, multiply 4, one-half the coefficient of x, by itself, making 16 and subtract the number 7, leaving 9. The square root of this added to or subtracted from 4, one-half the coefficient of x, gives 7 and 1 as the values of x. Since, however, 1 is not a [proper] number for an association, we say that x is 7, and this is the number of [43]men in the first[43] association. The second, therefore, will have 14 men and the third 28. It is clear, also, that the cube of 7 plus the sum of the numbers of men is 392, and that this is also the product of 14, the second number, and 28, the third.

[42] 1545 has *irrationales*; 1570 and 1663 have *alogi*. This difference in terminology occurs very frequently hereafter and is not further noted.

[43] 1663 omits these words.

CHAPTER VI

On Methods for Solving New Cases

1. After I had carefully considered all this, it seemed to me that it would be permissible to go still further and thus, following the example of the derivatives which had already been discovered—namely, the fourth power[1] and square equal to the number, the sixth power and cube equal to the number, and the other four—I set up the case of the eighth power, fourth power, and constant equal to each other in turn. Here the value of x is the fourth root of its value in the corresponding principal case. Thus if

$$x^2 + x = 12^{[2]}$$

and the value of x is 3, then in

$$x^8 + x^4 = 12,$$

its value will be $\sqrt[4]{3}$. And so I applied my mind to thinking about the remaining derivatives.

2. Thereafter I directed myself to other matters and it seemed to me proper to think about the nature of solutions. So I gave attention to the origin of the first conjunction (for so we call a *binomium*) and the first *apotome* (for so we call its *recisum*). It seemed that the two quantities — the [whole] number and the surd or root — are of different origins. If [a *binomium* or *apotome*] is squared, the [new] whole number is the [sum of] the squares of each of the parts and the [new] root is twice the product of the two parts. The irrational part of the cube, moreover, is made up of the root times the sum of three times the square of the [whole] number and the square of the root. Therefore, the ratio of the irrational part in the cube to the irrational part in the square is the same as that of thrice the square of

$x: 2 + \sqrt{3}$
$x^2: 7 + \sqrt{48}$
$x^3: 26 + \sqrt{675}$

[1] 1663 has x^2.
[2] 1663 has $x^2 + 2x = 12$.

the part which is a whole number plus the square of the part which is a root to twice the number. But the ratio of three times the square of the number to twice the number is one and a half times the number itself. And the ratio of the square of the root to twice the number is that which comes from dividing such a square by twice the number. Therefore, the [whole] ratio is one and a half times the number plus such a quotient. Hence, given x^2 with a certain coefficient [equal to x^3 and a constant], the irrational parts [of the squares and cube] will be equal and [the integral part of] the given number of squares will equal [the integral part of] the cube plus the constant. Thus, in the above case [where $x = 2 + \sqrt{3}$], divide 3, the square of the root, by 4. The quotient is $\frac{3}{4}$. Add 3 (which is equal to $1\frac{1}{2}$ times the [whole] number) to this, and it makes $3\frac{3}{4}$. I say that in this case

$$3\tfrac{3}{4}x^2 = x^3 + N,$$

and N is $\frac{1}{4}$.

At length, wishing to scrutinize the matter still more diligently, I assumed $10x^2$ to be equal to x^3 and a number, and I assumed the first part of the *binomium* (for thus, thanks to usage, I will call the conjunction) to be 3, for example, and I made y[3] the second part or root. The square of this is $9 + y^2$ (which is the [new whole] number) plus $6y$ (which is the root). But in the cube, as has been said, the irrational part is the irrational part [of the first power], that is y, times [the sum of] thrice the square of 3, which is 27, and the square of y, which is y^2. Hence $27y + y^3$ equals the irrational part of $10x^2$, which is $60y$. Hence we say that y^3 equals $33y$ and, dividing by y, that y^2 equals 33. Hence y equals $\sqrt{33}$ [and the number in

$$10x^2 = x^3 + N$$

is 96.]

Rule

From this the following very brief rule may be formulated: Add to the first number [in the *binomium*] one-half of itself and subtract the sum from the coefficient of x^2. Multiply the remainder by twice the first number, and the square root of the product is the second part of the *binomium*.[4]

[3] I have adopted y here instead of x, in accordance with modern usage, to avoid confusion. Cardano uses *positio*, but not uniformly so.

[4] I.e., if $x^3 + N = ax^2$, and if $x = s + y$, $y = \sqrt{2s(a - \frac{3}{2}s)}$.

For example,

$$x^3 + N = 12x^2$$

and the first part of the *binomium* is 5. Add one-half of 5 to 5, making $7\frac{1}{2}$. Subtract this from 12, leaving $4\frac{1}{2}$. Multiply $4\frac{1}{2}$ by 10, twice 5, the first number [of the *binomium*], making 45, the square root of which is the second part of the *binomium*. Hence, with x equal to $5 + \sqrt{45}$,

$$12x^2 = x^3 + 40.$$

By the same reasoning, I discovered that the constant of the equation, namely 40, is the difference between the product of the difference between the first number [of the binomium] and the coefficient of x^2 times the square of the first number and the product of [the difference between] three times the first number and the coefficient of x^2 times the square of the root.[5]

From here I went on to explore the nature of the cases involving a cube, square, first power, and constant, and I saw that if I should say

$$x^3 + 3x^2 = 14x + 20$$

and the value of x were assumed to be some known quantity, its first part would be a number — that is, a true quantity — and the other an irrational.[6] In this case, for example, it is $1 + \sqrt{5}$. It is clear, moreover, that the sum of the irrational parts of the cube and square is twice the coefficient of x^2 times the first part of the solution (or the coefficient of x^2 times twice this number) plus three times the square of this number plus the square of the irrational part [all three being multiplied by the irrational part].[7] It is the same in the case of a cube, square, and number [equal to the first power] and here it supplies the coefficient of x. Therefore, in either case, the rational part is such that the sum of twice the coefficient of x^2 [times the rational part] plus three times the square of the rational part plus the square of the other part will constitute the coefficient of x. If the rational part, or number, is negative, it is reasonable that the coefficient of x should be the difference between twice the coefficient of x^2 [times the integral

[5] Is this not a beautiful example of how much we depend on symbols for clarity? What Cardano is saying here is this: If $x^3 + N = ax^2$ and if $x = s + y$, $N = s^2(a - s) - y^2(3s - a)$.

[6] The meaning of this is quite clear, but the text seems somewhat garbled: *et ponatur quantitas quaedam intellecta, aestimatio rei, cuius prima pars sit numerus, secunda vero quantitas, alia pars irrationalis.*

[7] I.e., if $x^3 + bx^2 = ax + N$, and if $x = s + y$, $ay = (2bs + 3s^2 + y^2)y$.

part of the *apotome*] and three times the square of the part which is a whole number plus the square of the part which is a root.[8,9] For example, if

$$x^3 + 6x^2 + N = 30x,$$

and one part of the *apotome* is -2, we multiply 6, the coefficient of x^2, by 4, twice 2, making 24. To this we add 30, the coefficient of x, making 54. This ought to equal three times the square [of -2], which is 12, plus the square of the other part. Hence subtracting 12 from 54 leaves 42, and $\sqrt{42}$ is the first part of the *apotome*. Hence x equals $\sqrt{42} - 2$.

3. There is another method, which is called [the method of] similitude. It is four-fold: [1] From the nature of the equation, as when the [solution in the] case of the cube equal to the first power and number is derived from the [solution in the] case of the cube and first power equal to the number. [2] From an augmentation of solutions, and thus we discovered the nonuniversal case of the fourth power, first power, and number.[10] [3] By the conversion of equations into [others] equivalent in nature, as we will show hereafter. [4] By the method of proceeding[11] to solutions by the creation of cubes and squares, or by proportion (as half or double), or by addition and subtraction, these being three variant methods within a whole.[12]

4. There is also the way of transmutation by which I discovered many general rules before [their] demonstration and, among others, those of the cube equal to the square and number and of the cube and square equal to the number, as when we undertook to solve this problem: Find two numbers the sum of which is equal to the square of the second and the product of which is 8. By one route you arrive at

$$x^3 = x^2 + 8,$$

by another at

$$x^3 + 8x = 64.$$

[8] The text has "three times the square of the part which is a root plus the square of the part which is a whole number."

[9] I.e., if $x^3 + bx^2 = ax + N$ and if $x = -s + y$, then $a = 3s^2 + y^2 - 2bs$, ignoring the minus sign before s.

[10] 1663 can be read as saying either "eighth power, first power, and number" or "fourth power, square, first power, and number."

[11] 1545 and 1570 have *pro edendi*; 1663 has *procedendi.*

[12] 1545 has *tres enim sunt modi variandi in universum*; 1570 and 1663 have *tres enim sunt modi in universum.*

Having discovered the solution [for the latter], if you divide 8 by it, it will produce the solution for the other. From this in that case, I came to this thought: Translate problems that are, by some ingenuity, understood into terms of the unknown,[13] and the discovery of rules can have no end and, beyond this, general principles may be discovered from a single problem.

5. When, moreover, I understood that the rule that Niccolò Tartaglia handed to me had been discovered by him through a geometrical demonstration,[14] I thought that this would be the royal road to pursue in all cases. So I undertook to set down three highly useful propositions. A clear exposition of these will make it easy to understand others which will also be demonstrated. Here is the first of them.

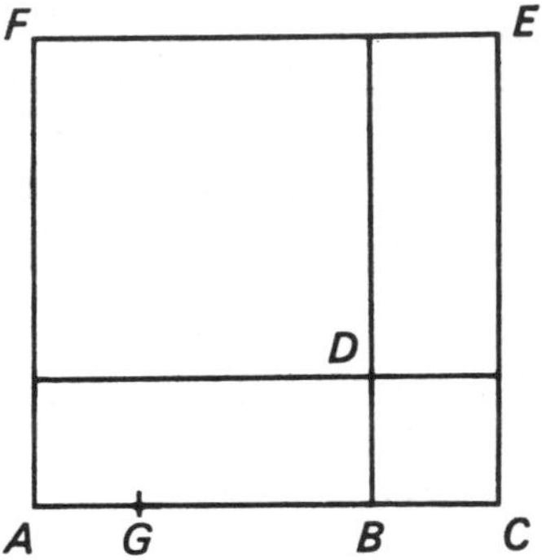

6. If a quantity is divided into two parts, the cube of the whole is equal to the cubes of the two parts plus three times the products of each and the square of the other.[15] Although this and the two which follow have been shown elsewhere by us in our seventh book on Euclid,[16] it is still proper to set it [the demonstration] out again so that nothing may be wanting in this work.

Let, therefore, AC be divided at the point B and let the cube of the whole be AE and let, furthermore, DA, DC, DF, and DE be distinct surfaces on its base. It is manifest from II, 4 of the *Elements* that CD is the square of BC,[17] that DF is the square of AB, and that each of the two rectangles AD and DE is the product of AB and BC. The total cube, therefore, is made from the line AC times the square AE and, therefore, from AC times the surfaces DA, DC, DE, and DF, the components of AE. Hence, since AC consists of AB and BC, the cube AE will consist of eight bodies, four of which are made from the

[13] *Quaestiones igitur alio ingenio cognitas ad ignotas transfer positiones.*

[14] Cf. Tartaglia's *Quesiti et Inventioni Diverse* (facsimile reprint of 1554 edition, Brescia, 1959), p. 121v.

[15] I.e., if $N = a + b$, $N^3 = a^3 + 3ab^2 + 3a^2b + b^3$.

[16] 1545 has *Quamvis hoc et reliqua duo quae sequuntur in 7° nostro super Euclidem libro ostensa sint*; 1570 and 1663 have *Quamvis hoc et reliqua duo sequuntur alibi a nobis in 7 Elem. Geom. ostensa sint.*

[17] 1663 has BD.

line AB times the surfaces DA, DC, DE, and DF, and the remaining four from the line BC times the same four surfaces. But from AB times FD comes the cube of AB, and from BC times CD the cube of BC. The cube AE, therefore, consists of the cube of AB, the cube of BC, the products of AB and DA, DC and DE, and the products of CB and DA, DF and DE. But the product of AB and CD is equal to the product of BC and DA, and the product of BC and DF is equal to the product of AB and AD, the bases and altitudes of these being the same. The parallelepipeds $AB \times AD$ and $AB \times DE$ are likewise equal. Similarly for $BC \times AD$ and $BC \times DE$, because DA and DE are equal surfaces, as is shown in I, 43 of the *Elements*. Hence the cube of AC consists of the cubes of AB and BC plus thrice $AB \times BC^2$ and thrice $BC \times AB^2$, as was to be proved.

7. From this comes a second [proposition], namely that $AB^3 + 3(AB \times BC^2)$ is greater than $BC^3 + 3(BC \times AB^2)$ by the cube of the difference between AB and BC. Now let AG equal BC, and the difference between AB and BC will be the line GB. It follows from the preceding, however, that

$$AB^3 = AG^3 + GB^3 + 3(AG \times GB^2) + 3(GB \times AG^2).$$

Hence

$$AB^3 + 3(AB \times BC^2) = AG^3 + GB^3 + 3(AG \times GB^2)^{18} + 3(GB \times AG^2) + 3(AB \times BC^2).^{19}$$

Truly AG^3 equals BC^3, and $3(BG \times AG^2)$ equals $3(BG \times BC^2)$, and $3(AG \times GB^2)$ equals $3(BC \times BG^2)$, since BC equals AG. Therefore

$$AB^3 + 3(AB \times BC^2) = BC^3 + BG^{3\,20} + 3(BG \times BC^2) + 3(BC \times BG^2) + 3(AB \times BC^2).^{21}$$

But $BG \times BC^2$ equals BC times the rectangle $BG \times 3BC$. Therefore $BG \times BC^2$ equals BC times the rectangle $BC \times 3BG$. For the same

[18] The text has $3(AB \times GB^2)$.

[19] 1663 has $AB^3 + 3(AB \times BC^2) = AG^3 + GB^3 + 3(AB \times GB^2) + 3(GB \times AG^2) + 3(AB \times GB^2) + 3(GB \times AG^2) + 3(AB \times BC^2)$, the fifth and sixth terms in the right-hand member of the equation being a repetition of the two that precede them.

[20] The text has BG.

[21] 1570 reads $AB^3 + 3(AB \times BC^2) = BC^3 + BG^{[3]} + 3(BG \times BC^2) + 3(BC \times BC^2) + 3(BG \times BC^2) + 3(BC \times BG^2) + 3(AB \times BC^2)$, the second and third terms in the right-hand member of the equation being (except for a misprint in $3(BC \times BC^2)$) a duplication of the two items which immediately follow it. The printer apparently inserted the correct material after proofreading but neglected to remove that which it replaced.

reason, $AB \times 3BC^2$ equals BC times the rectangle $AB \times 3BC$. Therefore,

$$AB^3 + 3(AB \times BC^2) = BG^3 + BC^3 + 3BC(BC \times AB) + 3BC(BC \times BG) + 3BC(BG^2).$$

But from II, 4 of the *Elements*, the rectangle $BC \times BA$ plus [the rectangle] $BC \times BG$ plus BG^2 equals AB^2. Therefore,

$$BG^3 + BC^3 + 3(AB^2 \times BC)^{22} = {}^{23}AB^3 + 3(AB \times BC^2)^{23}$$

Whence $AB^3 + 3(AB \times BC^2)$ is greater than $BC^3 + 3(BC \times AB^2)$ by the cube of the difference, BG.

Corollary 1. From this it is plain that, if BC is assumed to be negative, AB^3 will consist of $AC^3 + 3(AC \times BC^2) + (-BC^3)$[24] $+ (-3BC \times AC^2)$. For if BC were plus, the difference between $AC^3 + 3(AC \times BC^2)$ and $BC^3 + 3(BC \times AC^2)$ would be AB^3, as has been demonstrated. But BC being negative, as much has to be added as was the difference given when BC was plus. Therefore

$$AB^3 = AC^3 + 3(AC \times BC^2) - 3(BC \times AC^2) - BC^3.$$

And, likewise, if AB is assumed to be negative, BC^3 will consist of $AC^3 + 3(AC \times AB^2) - 3(AB \times AC^2) - AB^3$.

Corollary 2. In the same way, if AB is assumed to be negative, its cube is composed of $BC^3 + 3(BC \times AC^2) - AC^3 - 3(AC \times BC^2)$ for, as has been said, AB^3 is the difference of these parts [i.e., the difference between the first two and the second two] in the positive, from the first corollary. Therefore, subtracting the greater from the less gives the same result in the negative. But $-AB^3$ is equal numerically to $+AB^3$, just as $+27$ is $(+3)^3$ and -27 is $(-3)^3$. Therefore,

$$-AB^3 = BC^3 + 3(BC \times AC^2) - AC^3 - 3(AC \times BC^2).$$

8. From the first proposition, moreover, a third can be shown, which is that the ratio of $AB^3 + BC^3$ to $3(AB \times BC^2) + 3(BC \times AB^2)$ is the sum of the first and third minus the second of three proportional quantities in the proportion $AB:BC$ to three times the second of these.[25]

[22] 1570 and 1663 have $3(AB \times BC^2)$.

[23] 1570 and 1663 omit this material.

[24] *addito per m:.*

[25] 1545 has *ut trium linearum in proportione continua, AB et BC existentium aggregati primae et tertiae, detracta secunda, ad triplum secundae*; 1570 and 1663 have *ut aggregati primae et tertiae detracta secundam trium quantitatum analogarum in proportione ab ad bc ad triplum secundae earum.*

It follows from XI, 32 of the *Elements* that the ratio of AB^3 to the body $BC \times AB^2$[26] is as AB^2 to the surface AD [i.e., $AB \times BD$], and therefore, from VI, 1 of the *Elements* as AB to BC. By the same reasoning, the ratio of the parallelepiped $BC \times AB^2$ to the parallelepiped $AB \times BC^2$ is as AB to BC. Again, [that] of the parallelepiped $AB \times BC^2$ to BC^3 is as AB to BC. There are, therefore, four bodies — AB^3, the parallelepiped $BC \times AB^2$, the parallelepiped $AB \times BC^2$, and BC^3 — which are in the continued proportion of the lines AB and BC. For the sake of brevity, we will represent these bodies by the four letters H, K, L, M, so that H is AB^3, K is the parallelepiped $BC \times AB^2$, L is the parallelepiped $AB \times BC^2$, and M is BC^3. Therefore, since the ratio of M to L is that of L to K, as has been proved, K will be to L as H is to K, according to V, 24 of the *Elements*, and $K + M$[27] will be to L as $H + L$ is to K. Hence, according to the 12th proposition of the same, $H + K + L + M$ will be to $K + L$ as $H + L$ is to K. Therefore, according to the 19th of the same, $H + M$ will be to $K + L$ as $H + L - K$ is to K. Wherefore, according to the 22d of the same, $H + M$ will be to $3(K + L)$ as $H + L - K$ is to $3K$. But when H, K, and L are in the proportion of AB to BC, as has been proved, $AB^3 + BC^3$ will be to $3(AB \times BC^2) + 3(BC \times AB^2)$, according to V, 11 of the *Elements*, as the first and third of three proportional lines in the proportion AB and BC, subtracting the middle one, to three times the same middle.

Corollary 3. It is clear from this that the ratio $\frac{3(BC \times AB^2)}{3(AB \times BC^2)}$ is that of AB to BC[28] from V, 12 of the *Elements*.

Corollary 4. [It is also clear that] the ratio of $AB^3 + BC^3 + 2(BC \times AB^2) + [2](AB \times BC^2)$ to the remainder of the whole cube AC is as the three surfaces $DC + DA + DF$[29] to the surface DE or as three mean proportional quantities in the proportion AB to BC to their mean. [There are] many other [corollaries] which I omit for the sake of brevity.

[26] The text has $AC \times AB^2$.

[27] The text omits signs of addition here and for the remainder of this paragraph insofar as the H, K, L, M series is concerned.

[28] 1570 and 1663 have *ut abc* instead of *ut AB ad BC*.

[29] 1570 and 1663 have DE.

CHAPTER VII

On the Transformation of Equations

1. When a number and a middle power are equal to a higher power, the equation may be converted to one of the same two powers, with the same coefficients, equal to the number. As, if I say,

$$x^2 = 6x + 16$$

we can also say

$$x^2 + 6x = 16$$

and conversely. Then, having the solution for the first of these, we subtract from it the coefficient of x, which is 6, and we will have the solution for the second or, having the second, we add to it 6, the coefficient of x, and thus make the solution of the first. Truly, [however], no general rule can be given for the other powers.

2. There is a general rule that when a middle power is equated to a number and a higher power, it can be converted to another middle power as far removed from the number as the first was removed from the higher power. Thus, for example, if the cube and number are equal to the first power, the cube and number will be equal to the square but not with the same coefficient that the first power[1] had. The rule for arriving at the [new] middle term is this: Divide the higher of the two middle powers by the lower; take the root of the number in accordance with the nature of the highest power; reduce the result to the power that has emerged [from the preceding division]; and multiply the coefficient of the middle power which is closer to the highest [or] extreme power[2] [by the result] or divide the coefficient [of the term] which is closest to the number, and the product or quotient is the coefficient of the [new] middle power. As, if

$$x^3 + 16 = 6x^2,$$

[1] 1545 has *sed non sub rerum numero existentibus*; 1570 and 1663 omit *existentibus*.

[2] *multiplicabis numerum denominationis mediae proximioris maximae denominationi extremae.*

from what has been said

$$x^3 + 16 = ax.$$

We will look for the [new] coefficient thus: Divide x^2 by x, and x is the result. Take the cube root of 16, for x^3 is the highest power, and reduce this to the nature of x, since x is what emerged when x^2 was divided by x. This makes, therefore, $\sqrt[3]{16}$, since x neither increases nor diminishes it. Then multiply $\sqrt[3]{16}$ by 6, the coefficient of x^2, which is closer to the cube than to the number, and we get

$$\sqrt[3]{3456}x = x^3 + 16.$$

Another example,

$$x^3 + 8 = 18x.$$

You say, therefore,

$$x^3 + 8 = ax^2.$$

Divide x^2 by x, leaving x. Take the cube root of 8, since x^3 is the highest power, and this is 2. This cannot be reduced further, since x is the term which emerged. Let 2 be the divisor of 18, the coefficient of x, since x is closer to the number than to the cube, and 9 results as the coefficient of the square which is equal to the cube plus 8.

In the same way, if we say

$$x^4 + 64 = 10x^3,$$

the transformation into x's makes it

$$x^4 + 64 = ax.$$

Divide x^3 by x, leaving x^2. In accordance with the nature of the square of the square, take the fourth root of 64. This is $\sqrt{8}$. Square this according to the power that emerged, making 8. Multiply this by 10, the coefficient of x^3, since this is closer to the highest power, making 80 [the coefficient of] x. *Per contra*, divide 80 by 8 in order to have the coefficient of x^3.

3. The same procedure holds when the middle power and number are equal to the highest power or when the two higher powers are equal to the number, for we can make the transformation from one equation into the other by the same rule. For example,

$$x^3 = 9x + 10.$$

We say, therefore, that

$$x^3 + \sqrt[3]{72\tfrac{9}{10}}\,x^2 = 10.$$ [3]

And if

$$x^3 = 6x^2 + 16,$$

then

$$x^3 + \sqrt[3]{3456}\,x = 16.$$ [4]

And if

$$x^3 + 18x = 8,$$

then

$$x^3 = 9x^2 + 8.$$

And if

$$x^5 + 6x^3 = 80,$$

then

$$x^5 = ax^2 + 80.$$

In this case, divide x^3 by x^2 and x is the result. Take the fifth root of 80 and raise to the nature of x. It will remain $\sqrt[5]{80}$. Multiply this by 6, the coefficient of x^3, making $\sqrt[5]{622,080}$ for the coefficient of x^2. Hence,

$$x^5 \text{ [5]} = \sqrt[5]{622,080}\,x^2 + 80.$$ [6]

By the same reasoning, if

$$x^5 + 30x = 32,$$

then

$$x^5 = ax^4 + 32.$$

Divide x^4 by x, making x^3. Raise 2, the fifth root of 32, to the cube, making 8. Divide 24 [*sic*], the coefficient of x, by 8, making 3 the coefficient of x^4, and this plus 32 equals x^5.

4. To find the solution in individual cases, divide the square of the root of the constant of the equation, the root being of the same nature as the highest power, by the solution which you have. What results

[3] 1570 and 1663 have $\sqrt[3]{72,900}\,x^2$.

[4] 1570 and 1663 have 19.

[5] 1663 omits the *relatum* of *relatum primum*.

[6] 1663 has $\sqrt[5]{6222080}\,x^2$.

is the value of the converted case. For example, it has been pointed out that if

$$x^3 + 8 = 18x,$$

then

$$x^3 + 8 = 9x^2.$$

In the first equation, x equals 4 or $\sqrt{6} - 2$. I say that if you take the cube root of 8, which is 2, and multiply it by itself, making 4, and divide this by [either of] the first solutions, namely 4 or $\sqrt{6} - 2$, there will arise 1 and $\sqrt{24} + 4$[7] as the solutions for

$$x^3 + 8 = 9x^2.$$

And it has likewise been said that if

$$x^5 + 6x^3 = 80,$$

$$x^5 = \sqrt[5]{622{,}080}\,x^2 + 80.$$

In the first equation, the value of x is manifestly 2. Multiply, therefore, $\sqrt[5]{80}$ by itself, making $\sqrt[5]{6400}$. Divide by 2, the solution for

$$x^5 + 6x^3 = 80.$$

The result is $\sqrt[5]{200}$, the value when

$$x^5 = \sqrt[5]{622{,}080}\,x^2 + 80.$$

In order that all of these things may be more clearly understood, there are subjoined 24 transformations, from which others may be discerned. Here there are 12 transformations one way and an equal number in reverse. Thus, if the cube and square are equal to a number, this may be converted into the case of the cube equal to the first power and number. Contrariwise, if the cube is equal to the first power and a number, the cube and square will also equal the number.

$x^3 + ax^2 = N$	into	$x^3 = bx + N$
$x^3 = ax^2 + N$	,,	$x^3 + bx = N$
$x^3 + N = ax^2$	,,	$x^3 + N = bx$
$x^4 + ax^3 = N$	,,	$x^4 = bx + N$
$x^4 + N = ax^3$	,,	$x^4 + N = bx$
$x^4 = ax^3 + N$	,,	$x^4 + bx = N$

[7] The text has $\sqrt{24} - 4$.

$x^5 + ax^4 = N$	into	$x^5 = bx + N$
$x^5 = ax^4 + N$	,,	$x^5 + bx = N$
$x^5 + N = ax^4$	,,	$x^5 + N = bx$
$x^5 + ax^3 = N$	,,	$x^5 = bx^2$[8] $+ N$
$x^5 = ax^3 + N$	,,	$x^5 + bx^2 = N$
$x^5 + N = ax^3$	,,	$x^5 + N = bx^2$

Demonstration

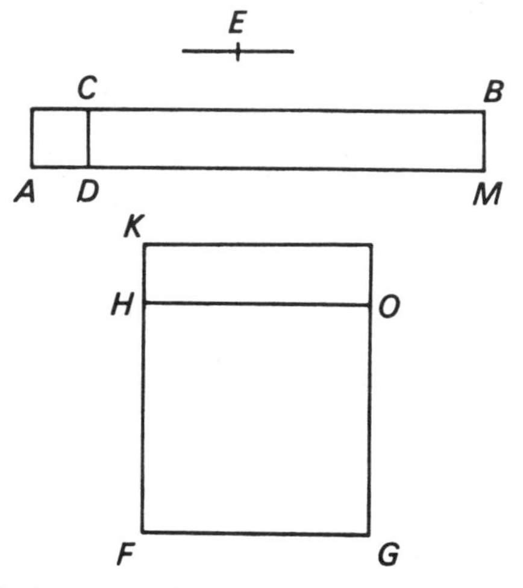

5. In order to demonstrate this method, assume, for example, that the parallelepiped *AB* is made up of *AC*, the cube, and *DB*, the number, and assume that the sum of these is equal to [a certain number of] squares of the line *AD*. Therefore, since this [i.e., the parallelepiped] is the product of *DC* and [the surface] *AB*, it will be the product of *AM* and DC^2. Therefore *AM* is the coefficient of x^2. Between *MD* and *DA* let there be continued proportionals, *E* being closer to *AD* and *FG* closer to *DM*. Let the square of *FG* be *GH*, and let the surface *GK* be equal to the product of *E* and *AM*. Complete the body *GK* with *FG* as the altitude. Then, according to VI, 16 of the *Elements*,[9]

$$AM:FK = FG:E.$$

Therefore, according to V, 11 of the same,

$$AM:FK = MD:FG \text{ or } FH.$$

But, from V, 19 of the *Elements*,

$$AM:FK = AD:KH$$

and, therefore, according to the 11th of the same,

$$MD:FH = AD:KH.$$

[8] 1570 and 1663 have x^4.

[9] The text has VI, 15.

Therefore, since

$$MD:FH = FH:E$$

and

$$FH:E = E:AD$$

and

$$E:AD = AD:KH,$$

these five lines — *MD*, *FH*, *E*, *AD*, and *HK* — are in continued proportion. Therefore, according to XI, 34[10] and VI, 17 of the *Elements*,

$$GH:AC = MD:HK.$$

Now both of these [ratios] are duplicates of the ratio *FH* and *AD*. Therefore, the product of *DM*, the fifth, and *AC*, the square of the second, is equal to the product of *KH*, the first, and *GH*, the square of the fourth. Therefore, the body *KO* is the proposed constant and it plus BC^3[11] is equal to as many x's as there are on the surface *GK*. But *GK* is equal to the surface $E \times AM$[12] and *E*, moreover, is the cube root of the given number *DB*, according to XI, 34 of the *Elements*,[13] and *AM* is the coefficient of x^2, as has been proposed. Therefore *GK*, the coefficient of x, is the product of the cube root of the constant and the coefficient of x^2, and the constant of the equation remains the same, namely the body *KO* [or][14] *BD*, which we have demonstrated to be equal to each other. It remains, accordingly, for us to determine the value of x, which is *AD* in one and *FG* in the other. These were assumed to be such that *E*, the cube root of the given constant, lies between them as a mean proportional. Therefore, according to VI, 16 of the *Elements*, the square of *E* divided by either of them yields the other. Thus we have proved the remaining part of the rule. Generally, but briefly, the conclusion is that in those [cases] which are of this order, one can be known from the other.

RULE

6. There is also another form of transformation — viz., allowing the coefficients of the powers to remain the same but changing the constants. The reasoning here is the same as in the other [cases]. The rule

[10] 1570 and 1663 have XI, 32.

[11] The text has BG^3.

[12] The text has $C \times AM$.

[13] This reference appears to be totally out of place here.

[14] The text has *et*.

is this: Take the root of the constant of the equation according to the nature of the middle power which you already have and adjust[15] it to the nature of the middle power which you wish to equal the extremes in the converted form, and this is the constant of the second equation.

For example, if I say

$$x^3 + 8 = 18x,$$

you know by the table given above, which relates to this rule, that it will change to

$$x^3 + N = bx^2.$$

And from the present rule it is clear that the coefficient of x^2 is equal to the coefficient of x. Therefore,

$$x^3 + N = 18x^2.$$

To find the number, then, use 8 — since x has no root — and square it, which makes 64 the constant of the [transformed] equation. You square it because the middle power into which the change is to be made is x^2.

By the same reasoning, if it were said that

$$x^4 + 8 = 12x,$$

this can be changed into

$$x^4 + N = bx^3.$$

To do this, take the cube of 8, thus making

$$x^4 + 512 = 12x^3.$$

Similarly, if it were said that

$$x^5 + 8 = 5x^3,$$

the change may be to

$$x^5 + N = 5x^2$$

according to the table or the rule. To derive the number, since the middle power in the equation is a cube, take the cube root of 8, which is 2,[16] and raise it to the nature of the square, since this is the middle power of the transformed equation. This makes 4. Hence

$$x^5 + 4 = 5x^2.$$

7. The same method holds when [the equation of] the number and

[15] *reduces multiplicando.*
[16] 1570 and 1663 omit this figure.

middle power equal to the highest power is changed to an equation of the [two] powers equal to the number.[17] For example, if we said

$$x^5 + 4x^3 = 64,$$

we would, because of the cube, take the cube root of 64, which is 4, and raise this to the square, the middle power into which the transformation is to be made, and we will have

$$x^5 = 4x^2 + 16.$$

And if

$$x^5 + 4x = 5,$$

since x has no root, we raise 5 to its fourth power, which is 625. Then we say that

$$x^5 = 4x^4 + 625.$$

8. The method of solution in [the case of] the middle power equal to the highest power and the number is this: Raise the solution which you have in accordance with the nature of the middle power into which the change is to be made, and subtract this from the coefficient of the middle power. The root of the remainder, taken according to the nature of the middle power from which the change was made, is the value of x.

For example, if

$$x^5 + 64 = 12x^3$$

we can say

$$x^5 + 16 = 12x^2.$$

The solution for the first equation is 2 and, since the middle power of the equation into which the change is being made is a square, we square 2, making 4. Subtract this from 12, the coefficient of x^3, leaving 8. We take the root of this remainder according to the nature of the middle power from which the transformation is being made. This is a cube. Therefore, the cube root of 8, which is 2, will also be the value of x in the second equation.

Another example: if

$$x^5 + 64 = 24x^2,$$

you know that it can be changed to

$$x^5 + 512 = 24x^3.$$

The solution of the first, moreover, is 2, the cube of which is 8 — for the middle power of the second equation is a cube. Subtract 8 from

[17] It will be noted that the examples that follow illustrate the reverse of the text.

24,[18] the coefficient of x^2, and 16 remains. Take the square root of this — this is according to the nature of the middle power of the first equation — which is 4. This is the solution for

$$x^5 + 512 = 24x^3.$$

9. Where[19] the middle power is on the side either of the number or of the highest power, having made the change in accordance with the seventh rule, reduce the solution which you have, as before, to the nature of the middle power [of the equation] for which you seek the solution and, if the middle power whose value you seek is joined to the number, add to it [i.e., to the reduced solution] the coefficient of the middle power or subtract the coefficient if the middle power is joined to the highest power. The root of the sum or difference of these having been taken according to the nature of the middle power [of the equation] whose value is known, the solution of the second equation will be found.

Example: if

$$x^5 = 3x^3 + 8,$$

and the known value of x is 2, let this equation be transformed, according to the seventh rule, into

$$x^5 + 3x^2 = 4.$$

Take the square of 2, therefore, [since this is the nature] of the middle power [in the equation] the solution of which we seek. This is 4. From this subtract 3, the coefficient of x^2, since it is to x^5 that x^2 is joined and not to the number. This leaves 1. The cube root of this — since the cube is the middle power of [the equation] the solution which is already known — is 1, and this is the value of x.

Again, let

$$x^5 = 7x^2 + 4,$$

and let the change be into

$$x^5 + 7x^3 = 8,$$

in accordance with the seventh rule, and let the solution for this, which is 1, be known. You wish the other [solution]. Therefore, square 1 [because the square is] the middle power [in the] unknown [equation]. This is 1. To this add 7, the coefficient of x^2, since [it is] the middle power [which is] unknown [and] which is x^2, [which] is on the side of

[18] 1570 has 42.

[19] 1545 and 1570 have *ubi*; 1663 has *quia*.

the number, namely 4 — and we will have 8. Take the cube root in accordance with the nature of the middle power [of the equation whose solution is] known. This is 2. Its cube root is the value of x when

$$x^5 = 7x^2 + 4.$$

Corollary. From this it is evident that, having one equation, one can always have another by the second, third, and fourth rules, or by the sixth, seventh, eighth, and ninth — general if [the original is] general, or particular if [the original is] particular. Knowing the rule for the cube and first power equal to the number, take

$$x^3 = 3x^2 + 10$$

as an example. According to the seventh rule, we will have

$$x^3 + 3x = \sqrt{10},$$

the solution of which is $\sqrt[3]{\sqrt{3\frac{1}{2}} + \sqrt{2\frac{1}{2}}} - \sqrt[3]{\sqrt{3\frac{1}{2}} - \sqrt{2\frac{1}{2}}}$. The square of this plus 3, the coefficient of x^2 (since the square is joined to the number) will be the solution for

$$x^3 = 3x^2 + 10.$$

It is this because the given middle power, which is x, has no square root of itself. [It was] thus, in general, that I first discovered the rule for the cube equal to the second power and the constant, as well as many other rules — the way of the duplicate.

Demonstration

10. And lest this be not clear on its own account, a demonstration of it follows. This one will stand for all. Let the cube *DF* plus *AB*, a number, be equal to *DG* — that is, to the body *DG* — a number of x's. Moreover, let *HL*, the [same] number of x's, be equal to the surface *DG* times the number [*AB*]. And let the product of *HK* and *KM* be equal to *AC*, [another] number, which is the square of *AB*. Hence the product of *HL* and *KM* will

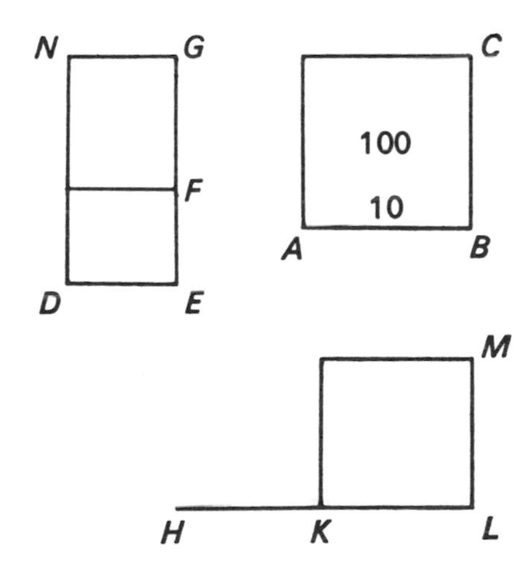

equal AC plus KL^3 and, likewise, DE times DG will equal DE^3 plus the number AB. DE, however, is the side of DF, and KL the side of KM. But HL equals DG. Since, therefore, HK times KM equals AC, and DE times FN equals AB, assuming that FN is the square root of KM and DE the square root of HK, I know[20] that if DE times FN equals AB, HK times KM will equal AC, for this is proved in the commentary of Theon on Euclid. Therefore, since the value of x is KL in one case and DE in the other, it follows that HL minus FD, which is equal to HK (since either is equal to DE^2), leaves KL, the value of x, as proposed.

11. There is another but dissimilar kind of transformation. When the fourth power equals the first power and constant and x is $\sqrt{5} + 2$, for example, then the fourth power and first power [with the same coefficient] will equal the same number, and x will be the *apotome* of the first solution, namely $\sqrt{5} - 2$, and *vice versa*.

12. [An equation] which consists of four terms, three of which are in continued proportion — say, the constant, x^2, and x^4 — and are equal either to x or x^3 can be transformed: Either[21] divide the coefficient of x by the square root of the constant and the coefficient of x^3 is the result; [or] multiply the coefficient of x^3 by the square root of the constant and there is produced the number of x's which are equal to x^4 and x^2 [22] and the constant with the same coefficients. For example, if

$$x^4 + 8x^2 + 64 = 10x^3,$$

multiply 8, the square root of 64, by 10, the coefficient of x^3 and [the result] will be

$$x^4 + 8x^2 + 64 = 80x.$$

Having the solution [for one equation], moreover, divide the square root of the constant by it and what results is [the solution for] the second. And if

$$x^4 + 8x^2 + 64\,[23] = 56x,$$

and x is 2,[24] then

$$x^4 + 8x^2 + 64 = 7x^3.$$

[20] The text has *nescio* (I do not know), apparently an error for *noscito* or something similar.

[21] 1545 and 1570 have *nam*; 1663 has *vel*.

[22] 1663 omits this last term.

[23] 1570 and 1663 have 46.

[24] The text has 4.

Divide 8, the square root of 64,[25] by 2, the first solution; this leaves 4 as the second solution, viz., that for

$$x^4 + 8x^2 + 64 = 7x^3.$$

13. There is also a transformation of equations consisting of three terms into equations of four terms. As an example, I will expound one rule: If we have an equation of the cube and number equal to the square, it can be changed into the cube and first power equal to the square and constant this way, the coefficient of x^2 remaining the same: Multiply half the coefficient of x^2 by itself and the result is the coefficient of x, which is on the same side as the cube. The [new] constant is always one-eighth the first constant; this is placed with the square. The solution always remains the same. For example,

$$x^3 + 16 = 14x^2.$$

Square 7, one-half of 14, making 49. Take one-eighth of 16, which is 2. You will then have

$$x^3 + 49x = 14x^2 + 2.$$

Another [example]:

$$x^3 + 40 = 8x^2.$$

Multiply 4, one-half of 8, by itself, and 16 is the result, the coefficient of x. Take one-eighth of 40; this is 5. Therefore,

$$x^3 + 16x = 8x^2 + 5$$

and, having derived the solution for one, you have it for the other. Since [the two solutions] are the same, no demonstration is necessary.

[25] 1570 and 1663 have 84.

CHAPTER VIII

Showing Generally the Solution for a Middle Power Equal to the Highest Power and Number

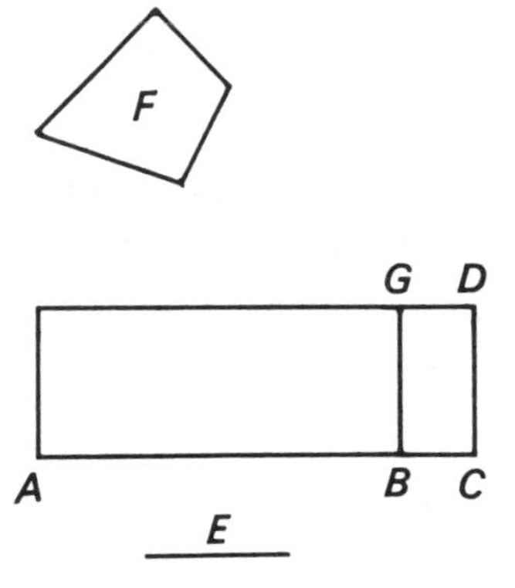

1. Let the sixth power and the number F be equal to a certain number of x's, and let AD be these x's and BD a portion of it. Having taken the root of this — that is, the fifth root — which is E and having multiplied it by AG, the remainder of the x's, there is produced F, the constant of the equation. I say that E is the value of x, for since, by supposition, E times AG is F and E times BD is E^6 [1] — because E is the fifth root of BD — the products of E and AG and of E and BD are equal to the product of E and AD. It follows, since AD is the number of x's, that the first power, with the value of E, equals the sixth power plus F, the number.

Rule

2. In accordance with this [demonstration], the rule can be formulated thus: When the highest [2] power and the number are equal to a middle power, divide the coefficient of the middle power into two parts such that one times the root of the other raised to the nature of the middle power — [the root] being taken according to the nature of the power obtained by dividing the highest power by the middle — yields the constant of the equation.[3] This root which was heretofore raised to the nature of the middle power is the value of x.

[1] The text has E^3.

[2] The text has *media* but the context makes clear that it should be *extrema*.

[3] I.e., if $x^h + N = ax^m$, let $a = b + c$ and let b and c be such that $b(\sqrt[h/m]{c})^m = N$. Then $x = \sqrt[h/m]{c}$.

Example:

$$10x = x^2 + 21.$$

Now, since x is exactly between x^2 and the number, it suffices to divide 10 into two parts the product of which is 21. These are 7 and 3, and either is the value of x.

Again,

$$10x = x^3 + 3.$$

Here x adjoins the number but not the cube, since x^2 lies between them. Hence we divide x^3 by x. The result is x^2 and we say, therefore: Make two parts of 10, such that one multiplied by the square root of the other gives 3. These will be 1 and 9 for 1 times 3, the square root of 9, equals 3. Accordingly, this same root 3, is the value of x.

Again,

$$10x^3 = x^4 + 64.$$

In this case, x^3 adjoins x^4 and is distant from the number by the intermediates x^2 and x. You say, therefore: Divide 10 into two parts [such that] one of them multiplied by the cube of the other produces 64. These parts will be 8 and 2, [the latter of] which must be cubed. Therefore 2 is the value of x [since] it is always properly the number which we manipulate that is the value of x.

Again (and this is an example of a fourth form[4]):

$$10x^3 = x^5 + 48.$$

Now since x^3 is distant from x^5 by x^4 lying between and from the number by the interposition of x^2 and x, divide x^5 by x^3 and x^2 will result. We now say: Divide 10, the coefficient of the middle power, into two parts one of which multiplied by the cube of the square root of the other produces 48, the constant of the equation. These parts will be 6 and 4, for from 6 times 8, the cube of 2, which is the square root of 4, comes 48. Hence this same 2, the square root of 4, is the value of x.

It is therefore evident that we always choose[5] our root according to the nature of the power by which the middle is contained in the

[4] The "fourth" because in the first example ($10x = x^2 + 21$), no root had to be taken or raised; in the second ($10x = x^3 + 3$), a root had to be taken but not raised; in the third ($10x^3 = x^4 + 64$), no root had to be taken but the number chosen had to be raised; and in this one ($10x^3 = x^5 + 48$), a root has to be both taken and raised.

[5] 1545 and 1663 have *sumimus*; 1570 has *suminus*.

greater power, and we raise this to the nature of the same middle power. He who knows how to do this can derive the rule, and he who knows how to derive the rule knows also how to do this.

3. It is evident that when the middle power is equal to the highest power and the constant, there are necessarily two solutions in all [cases] and most obviously so [in those] with a large constant.

CHAPTER IX

On a Second Unknown Quantity, Not Multiplied

Up to this point we have been treating of new discoveries quite generally. Now something must be said about certain individual types. It frequently happens that we must solve a given problem by using two unknown quantities. There follows an example of this which we could otherwise explain only with difficulty.

Three men had some money. The first man with half the others' would have had 32 *aurei*; the second with one-third the others', 28 *aurei*; and the third with one-fourth the others', 31 *aurei*. How much had each?

We let p be the first [man's share] and q the second's.[1] The third's, therefore, is $31 - (\frac{1}{4}p + \frac{1}{4}q)$. You see, accordingly, that the first has a quantity which, if you add half the second's and third's to it, is 32 *aurei*. For himself, then, he has $32 - \frac{1}{2}q - 15\frac{1}{2} + \frac{1}{8}q + \frac{1}{8}p$. Therefore he will have $16\frac{1}{2} - \frac{3}{8}q + \frac{1}{8}p$. Moreover, this is equal to p. Hence

$$\frac{7}{8}p + \frac{3}{8}q = 16\frac{1}{2}$$

which, being expressed in whole numbers, becomes

$$7p + 3q = 132.$$

Again, we see that the second man has an amount which is 28 *aurei* if to it is added one-third the sum of the first and third [shares] — that is, $\frac{1}{3}p + 10\frac{1}{3} - \frac{1}{12}p - \frac{1}{12}q$, or $\frac{1}{4}p + 10\frac{1}{3} - \frac{1}{12}q$. Subtract [this] from 28 and there remain $17\frac{2}{3} + \frac{1}{12}q - \frac{1}{4}p$, and so much is the second man's [share]. He, however, is supposed to have q, and therefore

$$q = \frac{1}{12}q + 17\frac{2}{3}\text{[2]} - \frac{1}{4}p.$$

[1] I have used p and q here, rather than the usual x and y, to come a bit closer to Cardano's *positio* and *quantitas*, these being the terms he employs a little later on in the problem.

[2] 1570 and 1663 have $37\frac{2}{3}$.

Subtract $\frac{1}{12}q$ from each side [of this equation] and transfer the negative to the opposite side. Then

$$\frac{11}{12}q + \frac{1}{4}p = 17\frac{2}{3},$$

wherefore, multiplying all terms by 12,

$$11q^{3} + 3p = 212.$$

Now raise whichever of these [two equations] you like to equality with the other with respect to the coefficient of either p or q. Thus you may decide that you wish, by some method, that there should be $7p$ in the equation

$$3p + 11q = 212.$$

Then, by the rule for four proportional quantities, there will be

$$[7p +]\ 25\frac{2}{3}q = 494\frac{2}{3}.$$

You will therefore have, as you see,

$$7p^{4} + 3q = 132$$

and

$$7p + 25\frac{2}{3}q^{5} = 494\frac{2}{3}.$$

Hence, since $7p$ is the same in both, in both the difference between the coefficients of q, namely $22\frac{2}{3}$, will equal the difference between the numbers, which is $362\frac{2}{3}$. Divide, therefore, as in the [case of a] simple unknown, according to the third chapter, $362\frac{2}{3}$ by $22\frac{2}{3}$; 16 results as the value of q and this is the second [man's share].

Again, assume the first to be p. The second, now, will be 16. Let the third be q. Since the second plus one-third the [sum of the] first and third is 28 [and] is [itself] 16,

$$\frac{1}{3}p + \frac{1}{3}q = 12,$$

[12 being] the difference between 28 and 16, and therefore

$$p^{6} + q = 36.$$

And,[7] further, since the first [share] with half the others is 32, half the others must be $8 + \frac{1}{2}q$. Therefore,

$$p + 8 + \frac{1}{2}q = 32.$$

[3] 1570 and 1663 have $1q$.
[4] 1570 and 1663 omit the 7.
[5] 1663 has $15\frac{2}{3}q$.
[6] 1570 and 1663 have $2p$.
[7] 1545 has *at*; 1570 and 1663 have *ad*.

Accordingly, subtracting 8 gives

$$p + \tfrac{1}{2}q = 24,$$

and since

$$p + q = 36,$$

the difference between 24 and 36, which is 12, equals $\frac{1}{2}q$. Therefore, following the method of the third chapter, divide 12 by $\frac{1}{2}$ and 24 results as the value of q or the number of *aurei* the third [man has]. Then, since it appears that the second has 16 and the third 24, and that the first, moreover, has 32, including half the second and third, subtract 20, one-half [the sum of] the second and third, from 32 and 12 remains as the number for the first [man's share]. Therefore the first [has] 12 *aurei*; the second, 16; and the third, 24. The operation is long, yet clear and easy. One term must always be reduced [so as] to [have] the same coefficient [in both equations], and then the difference between the numbers is necessarily the difference between the [coefficients of the] other term, as you saw twice in this example.

Another example: One man said to another, give me one-third of what you have and three [*aurei*] more and I will have three times what you still have. And the second said to the first, give me half of yours and two [*aurei*] more and what you then have will be one-ninth of all that I have. We let the first have p and the second q. Since, therefore, giving $\frac{1}{3}[q] + 3$ to the first leaves the second with $\frac{2}{3}q - 3$, and this is one-third the first, which is $p + \frac{1}{3}q$[8] $+ 3$, multiply the quantity $\frac{2}{3}q - 3$ by 3 — which will make $2q - 9$ — and this will be equal to $p + \frac{1}{3}q + 3$. Hence, by transferring what is negative to the other side [of the equation] there results

$$p + 12 = 1\tfrac{2}{3}q.$$

Again, since it was said that, if the first gave $\frac{1}{2}p + 2$ to the second, the remainder, namely $\frac{1}{2}p - 2$, will equal one-ninth the second's, that is $q + \frac{1}{2}p + 2$, therefore, by multiplying this residue by 9,

$$4\tfrac{1}{2}p - 18 = q + \tfrac{1}{2}p + 2$$

results; and by transferring to the other side the negative [quantities] and collecting similars, we will have

$$4p = q + 20.$$

[8] 1570 and 1663 have $\frac{2}{3}q$.

We also have

$$p + 12 = 1\tfrac{2}{3}q.$$

Now reduce the parts to equality in one term by multiplying

$$p + 12 = 1\tfrac{2}{3}q$$

by 4, making

$$4p + 48 = 6\tfrac{2}{3}q$$

and compare this, as you see in the table, with

$$4p = q + 20.$$

$$4p + 48 = 6\tfrac{2}{3}q$$
$$4p = 20 + 1q$$

$$4p + 68 + 1q = 4p + 6\tfrac{2}{3}q$$

$$5\tfrac{2}{3}q = 68$$

$$5q = 36 + 3p$$
$$5q + 100 = 20p$$

$$5q + 20p = 5q + 3p + 136$$

$$17p = 136$$

Similarly, by the same reasoning, reduce the coefficients of q to equality [and] you will have

$$5q = 36 + 3p$$
$$5q + 100 = 20p$$[9]

In either case, transfer in turn, according to the rule that if equals are added to equals the sums are equal, and you will have

$$4p + 68 + q = 4p + 6\tfrac{2}{3}q$$

and then, by subtracting similar terms, there will remain $5\tfrac{2}{3}q$ equal to 68. Accordingly, divide 68 by $5\tfrac{2}{3}$, and 12 results as the value of q and this is what the second man has.

By the same method, in the second equation, transfer unlike parts, saying if

$$[5]q = 36 + 3p$$[10]

and

$$5q + 100 = 20p,$$

therefore

$$5q + 20p = 5q + 3p + 136.$$[11]

Then, having subtracted likes, there will be left $17p$ equal to 136 and hence, by dividing 136 by 17, the result will be 8 as the solution for p or the first man's share. Thus the first has 8, the second 12. And

[9] 1570 and 1663 have $10p$.

[10] The text has $1q$, and 1570 and 1663 have $q = 39 + 2p$.

[11] 1570 and 1663 have $5q + 10p + 3p = 5q + 3p + 136$.

though these might be solved by other methods, yet this one is the better and clearer as it permits the whole problem to be solved at one time, as[12] thc first example by itself demonstrates.[13]

[12] 1545 and 1570 have *&* *si*; 1663 has *etsi*.

[13] 1570 and 1663 have at this point the following additional material:

A third quite suitable example: Find three quantities the first plus the second of which is one and one-half times the first and third and the first plus the third of which is one and one-half times the second and third. Let the third be 1, the second p and the first q. Or, better still, let the third be p and the second q. Then the sum of the first and third will be $1\frac{1}{2}p + 1\frac{1}{2}q$. Subtracting the third leaves the first as $\frac{1}{2}p + 1\frac{1}{2}q$. Similarly, since the sum of the first and second is one and one-half times the sum of the first and third, the sum of the first and second will be $2\frac{1}{4}p + 2\frac{1}{4}q$. And since the second is q, the first will be the remainder, or $2\frac{1}{4}p + 1\frac{1}{4}q$. The first quantity, from the first operation, was $\frac{1}{2}p + 1\frac{1}{2}q$ and, from the second operation, $2\frac{1}{4}p + 1\frac{1}{4}q$. These are equal to each other, according to the first axiom of Euclid. Again, in accordance with the third of the same, subtracting $\frac{1}{2}p$ and $1\frac{1}{4}q$ from both leaves $1\frac{3}{4}p$ equal to $\frac{1}{4}q$. Hence q equals $7p$. Assuming, therefore, that the third, which is $1p$, is 1, the second, which is $1q$, will be 7, and since their sum is 8 and the sum of the first and third is one and one-half times this, it will be 12 and since [the third] is 1, the first will be 11. Therefore the quantities are first 11, second 7, third 1, and the sums are 18, 12, and 8 in ratios of one and one-half as was proposed.

By the first method you would arrive at

$$q = 1\tfrac{1}{2}p + \tfrac{1}{2}$$

and

$$p = \tfrac{1}{2}q + 1\tfrac{1}{2}.$$

Then, doubling [the latter]

$$2p = q + 3.$$

But we have already shown that

$$q = 1\tfrac{1}{2}p \,[+]\, \tfrac{1}{2}.$$

Hence

$$2p = 1\tfrac{1}{2}p +^{14} 3\tfrac{1}{2}.$$

Therefore

$$\tfrac{1}{2}p = 3\tfrac{1}{2}$$

and

$$p = 7.$$

Accordingly, when

$$q = 1\tfrac{1}{2}p + \tfrac{1}{2}$$

and

$$p = \tfrac{1}{2}q + 1\tfrac{1}{2}$$

[we will have]

$$q = \tfrac{3}{4}q + 2\tfrac{3}{4}.$$

Therefore

$$\tfrac{1}{4}q = 2\tfrac{3}{4}$$

and q equals 11. This is a very pretty method since we are working with three quantities.

[14] The text has *d:* instead of *p:*.

CHAPTER X

On a Second Unknown Quantity, Multiplied

1.[1] When two unknown quantities are multiplied or squared, they take four forms, most of which have three members.

DEMONSTRATION

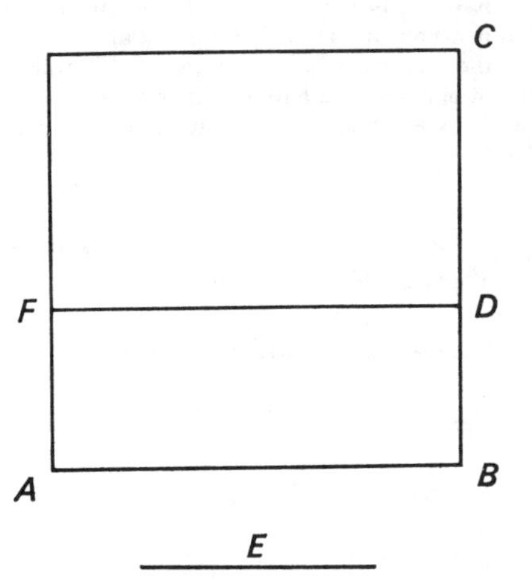

The first is when the square of one and the quantities themselves are compared. First, therefore, as an example, let the square AC, whose side is AB, be equal to $2AB + 5E$. Assume that BD is equal to the coefficient of x, namely 2, and AD will be equal to $2AB$. Therefore, CF is equal to $5E$. Hence, according to VI, 16[2] of the *Elements*,

$$AB:E = 5:CD.$$

AB, however, is x, CD is $x - 2$, and 5 is a known number. Therefore, the rule is:

RULE

Let x be what you please. Multiply it by itself minus the coefficient of x and divide the product by the coefficient of y. The quotient is the value of y.[3]

[1] 1570 and 1663 omit many of the section numbers in this chapter.

[2] The text has VI, 15.

[3] I.e., if $x^2 = ax + by$, and if x is given, $y = \dfrac{x(x - a)}{b}$.

For example, assume that x is 7. Multiply it by itself minus 2, since

$$[x^2] = 2x + 5y.$$

The result is 35. Divide 35 by 5, the coefficient of y. The quotient is 7, the value of y. And if we let x be 10, we multiply it by $10 - 2$ — that is, by 8 — and 80 is the result. Therefore, by dividing 80 by 5, we get 16, the value of the second unknown.

But if we assume that y is the known, we multiply it by its own coefficient and to the product we add the square of half the coefficient of x. The square root of the whole plus half the coefficient of x is the value of x.[4]

Example: Let y be 16. Multiplied by 5 it equals 80. If 1, the square of half the coefficient of x is added, it equals 81, the square root of which is 9. To this add half the coefficient of x, making 10, the value of x.

Demonstration

2. Again, as an example, let

$$10AB = AB^2 + 7E,$$

and let the square of AB be the surface AC and let BD be 10. Then $7E$ is equal to the surface FD and, as in the preceding,

$$AB:E = 7:CD.$$

Therefore, the rule for a first power equal to its square and a second unknown is —

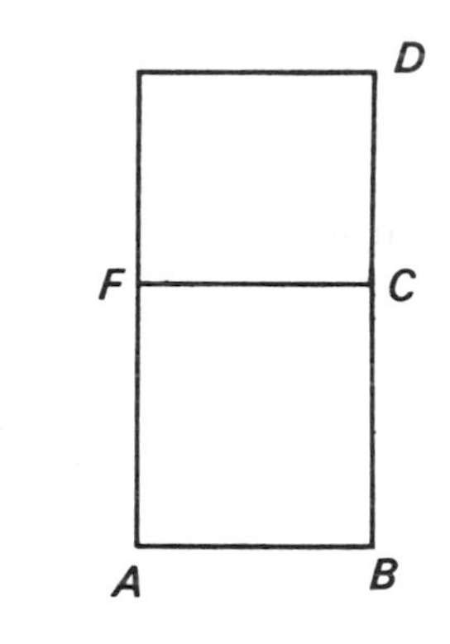

Rule

Assume x to be whatever you wish. Subtract it from the coefficient of x, and multiply it by the remainder. Divide the product by the coefficient of y. The result is the value of y.[5]

Example: Let x be 8 in this case. Subtract from 10, the coefficient of x, and 2 will remain. Multiply this by 8, producing 16. Divide by 7, the coefficient of y, and $2\frac{2}{7}$, the value of y, is the result.

But if it is the second quantity [that is assumed to be] known, multiply it by its own coefficient and subtract the product from the

[4] I.e., if $x^2 = ax + by$, and if y is given, $x = \sqrt{by + (a/2)^2} + a/2$.

[5] I.e., if $ax = x^2 + by$, and x is given, $y = \dfrac{x(a - x)}{b}$.

square of one-half the coefficient of x, and the root of the remainder, added to or subtracted from one-half the coefficient of x, is the value of x.[6]

Example: Let y be $2\frac{2}{7}$. Multiply this by 7, the coefficient of y, and 16 results. Subtract this number from 25, the square of one-half of 10, the coefficient of x, [7]and 9 remains, the square root of which added to or subtracted from 5, one-half of 10, the coefficient of x,[7] gives either 8 or 2 as the value of x.

Demonstration

3. Again, let there be a number of E's equal to the square of AB, which is AC, and to a number of AB's, which is the surface FD. Taking, therefore, AB first, the coefficient of E second, E third, and BD fourth, the ratio of AB to E will be that of the coefficient of E to BD. Whence comes the rule for a second unknown equal to a first unknown and its square —

Rule

Let x be whatever you please. Multiply it by the sum of itself and its coefficient and divide the product by the coefficient of y. The result is the value of y.[8]

For example,

$$5y = 7x + x^2.$$

Let x equal 3. We say, therefore, multiply 3 by 10 — the sum of 3, the [assumed] value of x, and 7, the coefficient of x — making 30. Divide this by 5, the coefficient of y, and 6 is the quotient and the value of y.

But if the second quantity is the known, we multiply it by its coefficient and to the product we add the square of one-half the coefficient of x, and the square root of the whole minus half the coefficient of x is the value of x.[9]

Example: Let 6 be the value of y when

$$5y = 7x + x^2.$$

Multiply 6, the value of y, by 5, the coefficient of y, making 30. To this add the square of $3\frac{1}{2}$, one-half of 7, the coefficient of x, and $42\frac{1}{4}$

[6] I.e., if $ax = x^2 + by$, and y is given, $x = a/2 \pm \sqrt{(a/2)^2 - by}$.

[7] 1663 has this passage twice.

[8] I.e., if $by = ax + x^2$, and x is given, $y = \dfrac{x(x + a)}{b}$.

[9] I.e., if $by = ax + x^2$, and if y is given, $x = \sqrt{by + (a/2)^2} - (a/2)$.

[results]. From the root of this, which is $6\frac{1}{2}$, subtract $3\frac{1}{2}$, half the coefficient of x, leaving 3 as the value of x.

Note. We have been using [two] unknowns, and no relation has been assumed between the two numbers at the beginning, either by way of addition or subtraction, multiplication, division, ratio, or root — for numbers may be related in these five ways. But if one [of these relations] exists, there is no necessity for [using] a second unknown, for the problem can be solved by one unknown.

Demonstration

4. If the product of x and y is compared to x and y, there will arise two sorts [of equations], for either the product will equal x and y, or x will equal the product and y. Let, therefore,[10] x be AB; y, AC; the coefficient of y, AD; and the coefficient of x, AE. Then, by hypothesis, the two surfaces DC and BE are equal to AF. AF, however, is equal to the four surfaces GA, GB, GC, and GF. Hence these four surfaces are equal to the surfaces DC and BE. Subtracting the three surfaces, GA, GB, and GC [from both] alike [i.e., from AF and from $DC + BE$], the other, GA, will be left equal to GF. Therefore, according to VI, 16[11] of the *Elements*,

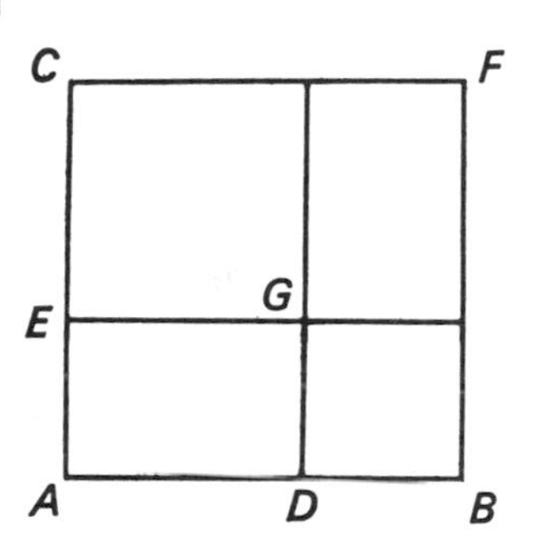

$$AD:DB = CE:EA.$$

The ratio, therefore of the coefficient of y to the remainder of x[, having subtracted the coefficient of y,] is that of the remainder of y, having subtracted the coefficient of x, to the coefficient of x. Accordingly, this will be the rule:

Rule

If x is known, we subtract from it the coefficient of y, and divide the product of the coefficients of x and y by the remainder. The quotient is added to the coefficient of x, and the total is y.[12]

[10] 1545 has *et*; 1570 and 1663 have *igitur*.

[11] The text has VI, 15.

[12] I.e., if $xy = ax + by$, and if x is given, $y = \frac{ab}{x - b} + a$.

For example, let

$$10x + 12y = xy,$$

and let y be 18. Then, conversely [— since here y is known, not x as in the stated rule —], subtract 10, the coefficient of x, from 18, the value of y, and 8 remains. Divide 120, which is the product of 10, the coefficient of x, and 12, the coefficient of y, by this. The result is 15, which added to 12, the coefficient of y, makes x equal to 27, whence $10x$ is 270 and $12y$ is 216. The sum of these, 486, is the product of 18, the value of y, and 27, the value of x. We have given this example of the converse of the rule in order that you may notice that the reasoning is one and the same [in both cases].

But if the product is known, divide it by the coefficient of y (if this is less than the coefficient of x) or by the coefficient of x (if this is less than the coefficient of y), and multiply half the quotient by itself. From this subtract that which results when the product of the greater coefficient and xy is divided by the smaller coefficient — the coefficient of x will be either greater or less — and the square root of the remainder added to or subtracted from that which was squared gives the value of y or of x, as the case may be — whichever is represented [as having] the smaller coefficient.[13] Then divide the product by this, and what results is assuredly the greater number.

For example:

$$2x + 6y = xy.$$

Let xy, for instance, equal 64. Then divide 64 by 2, which is less than 6. The result is 32, half of which is 16. Square this, making 256. Subtract from this 192, the result of dividing 384, the product of 6 and 64, by 2. The remainder is 64, the square root of which is 8. This added to or subtracted from 16, the number which was squared, gives the value of x as 8 or 24. Therefore, if x equals 8, y will likewise equal 8, for dividing 64 [14]by 8 leaves 8; and if x equals 24, y will equal $2\frac{2}{3}$, [the result of] dividing 64[14] by 24. In either case, $2x$ plus $6y$ will equal 64, the value of xy.

[13] I.e., if $ax + by = xy$, and if xy is given, and if $a > b$,

$$y = xy/2b \pm \sqrt{(xy/2b)^2 - axy/b}.$$

If, however, $a < b$,

$$x = xy/2a \pm \sqrt{(xy/2a)^2 - bxy/a}.$$

[14] 1663 omits this passage.

Demonstration

5. But if one side [i.e., one linear quantity] is equal to the product of this one and the other, plus the other side, let the first be AB and the second AE. Then the coefficient of the side AB is the surface AC. Therefore, EF results, by supposition, from the multiplication of AE by its own coefficient, which is $AB \times EC$. The ratio, therefore, of AB to AE [is] that of the coefficient of AE to EC. Then EC[15] is the difference between AE, or y, and AC, the coefficient of x. Therefore, the rule is as follows:

Rule

If

$$ax = xy + by$$

and y is known, we subtract this from the coefficient of x. Then we multiply y by its coefficient and divide the product by the remainder. The quotient is the value of x.[16]

For example,

$$10x = xy + 4y.$$

Let y equal 8. I subtract 8 from 10, and 2 remains. Then I multiply 8, the value of y, by 4, its coefficient, and the product is 32. This I divide by 2, the [above] remainder, and 16 results as the value of x. If the first subtraction cannot be made the case can be solved by no true number.

If, however, not y but x is the known, since the product of AB and AC is the same as AE times the sum of AB and the coefficient of AE, we divide the product of the coefficient of x and the value of x by the sum of x and the coefficient of y. The quotient is the value of y.[17]

For example,

$$10x = xy + 4y$$

and x is 16. I multiply 16, or x, by 10, the coefficient of x, producing 160. This I divide by 20, the sum of 4, the coefficient of y, and 16, the value of x. The quotient is 8, which is the value of y.

If it is xy that is known, multiply this by the coefficient of y and divide the product by the coefficient of x. To the quotient[18] add the square of

[15] 1663 has *EO*.

[16] I.e., if $ax = xy + by$, and if y is given, $x = \frac{by}{a - y}$.

[17] I.e., if $ax = xy + by$, and if x is given, $y = \frac{ax}{x + b}$.

[18] 1545 and 1570 have *cui exeunti*; 1663 has *exeunti*.

one-half that which results from dividing xy by the coefficient of x. The square root of the whole, plus the half that was squared in the first place, is the value of x.[19]

For example,

$$4x = 5y + xy.$$

Let xy be 45. I multiply 45 by 5, the coefficient of y, making 225. I divide this by 4, the coefficient of x, and $56\frac{1}{4}$ results. To this I add $31\frac{41}{64}$, the square of $5\frac{5}{8}$, half the result of dividing 45 by 4, and the sum is $87\frac{57}{64}$. The square root of this is $9\frac{3}{8}$ which, if it is added to $5\frac{5}{8}$, one-half the quotient from the [foregoing] division, gives 15 as the value of x.

DEMONSTRATION

6. When x^2, xy, and x are equated in turn, they take three forms. The first is when x^2 equals xy and the first unknown. Let x be AB,[20] the square of which is AC; let BF be y; and let AD by xy. The coefficient of xy will then be the number of times BF is contained in BD. DC will therefore be[21] the coefficient of x. Since, then, BC is equal to AB and CD is the coefficient of x, subtracting the coefficient of x from the value of x will leave BD, the product of the coefficient of xy and y. Hence the rule:

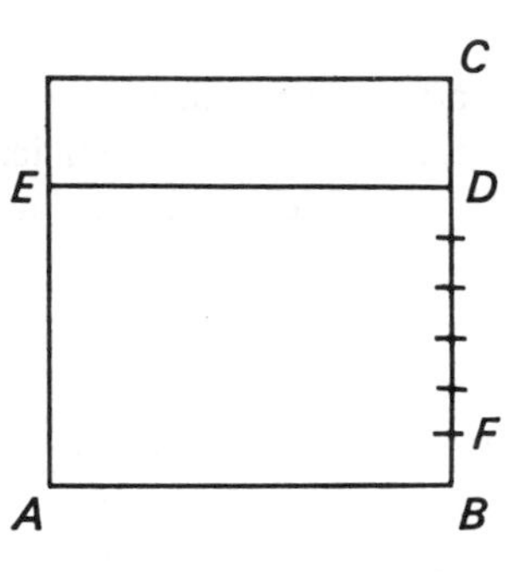

RULE

When x^2 is equal to the first power plus xy, if x is known, we subtract its coefficient from it, divide the remainder by the coefficient of xy, and [thus] produce the value of y.[22]

For example,

$$10x + 4xy = x^2.$$

X equals 30. I subtract 10 from 30, leaving 20. This I divide by 4, the coefficient of xy, and 5 is the quotient, the value of y.

But if it is y that is known, we multiply this by the coefficient of xy

[19] I.e., if $ax = xy + by$, and if xy is given, $x = \sqrt{\frac{bxy}{a} + \left(\frac{xy}{2a}\right)^2} + \frac{xy}{2a}$.

[20] 1545 and 1570 have *et sit AB res*; 1663 has *ut sit AB res*.

[21] 1545 has *DC igitur erit*; 1570 and 1663 have *DC igitur exit*.

[22] I.e., if $x^2 = ax + cxy$, and if x is given, $y = \frac{x - a}{c}$.

and to the product we add the coefficient of x and the value of x will derive therefrom.[23]

Example:

$$10x + 4xy = x^2.$$

Y equals 7. We multiply 7 by 4, the coefficient of xy,[24] making 28, to which we add 10, the coefficient of x, making 38 the value of x.

If it is xy that is known, we multiply this by the coefficient of xy[25] and to this we add the square of one-half the coefficient of x, and the square root of the whole, with half the coefficient of x further added [to it], is the value of x.[26]

For example,

$$x^2 = 10x + 4xy.$$

XY equals 50. We multiply 50 by 4, its own coefficient — that is, by the coefficient of xy — making 200, to which we add 25, the square of one-half of 10, the coefficient of x, making 225. To the square root of this add 5, one-half of the coefficient of x. This is 20, the value of x. Then by dividing 50, the product of x and y, $2\frac{1}{2}$ results as the value of y.

Demonstration

7. If xy equals the square of x plus a number of x's, we let x be AB, y be BC, and xy be AC and, because of this, DC will necessarily be the coefficient of x and AD the sum of the squares. Therefore, subtracting DC from BC leaves BD which, divided by the coefficient of x^2, will produce BF equal to AB. Therefore, the rule is:

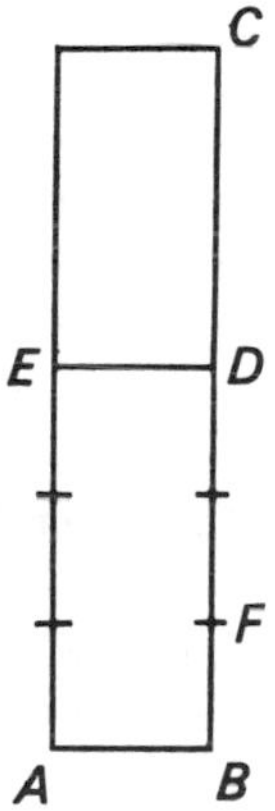

C

E D

F

A B

Rule

When xy equals the second power plus a number of x's, and x is known, we multiply this by the coefficient of x^2 and to the product we add the coefficient of x and [thus] will arise the value of y.[27]

For example,

$$xy = 6x^2 + 10x.$$

X is 4. Multiply 4 by the coefficient of x^2, 6, making 24. To this add 10, the coefficient of x, making 34, the value of y.

But if y is the known, we subtract the coefficient of x from it and

[23] I.e., if $x^2 = ax + cxy$, and if y is given, $x = cy + a$.

[24] The text has "the coefficient of y."

[25] 1663 has "the coefficient of x."

[26] I.e., if $x^2 = ax + cxy$, and if xy is given, $x = \sqrt{cxy + (a/2)^2} + a/2$.

[27] I.e., if $xy = ax + cx^2$, and if x is given, $y = cx + a$.

divide the remainder by the coefficient of x^2. The result is the value of x.[28]

For example,

$$xy = 6x^2 + 10x.$$

Y equals 34. I subtract 10 from 34 and 24 is left, which I divide by 6, the coefficient of x^2, producing 4, the value of x.

If it is xy that is known, we divide this by the coefficient of x^2 and to the quotient we add the square of one-half that which results from dividing the coefficient of x by the coefficient of x^2. The square root of the sum, after this same half has been subtracted, will be the value of x.[29]

For example,

$$xy = 6x^2 + 60x.$$

XY equals 1200. Divide 1200 by 6, the coefficient of x^2, and 200 results. To this add 25, the square of 5, half the result of dividing 60, the coefficient of x, by 6, the coefficient of x^2, making 225. From the square root of this, which is 15, subtract 5, the same half of the quotient, which leaves 10 for the value of x. Then, dividing 1200, which is the value of xy, by this, 120 will be produced as the value of y.

Demonstration

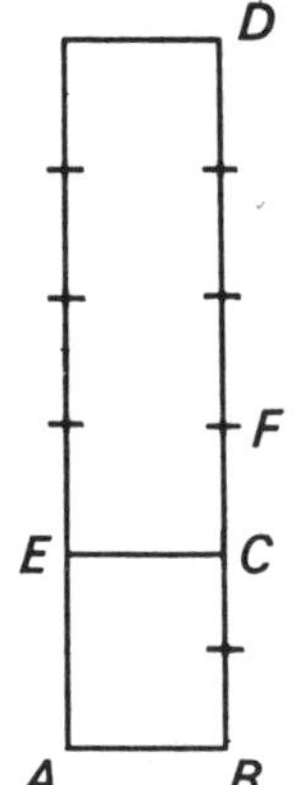

8. But if a number of x's are equal to the square of x plus xy (for indeed it is always permissible to reduce the square or xy to [a coefficient of] 1 by dividing all terms alike, as has been said in the general rules), we assume x to be AB, its square AC, and its coefficient BD.[30] Therefore ED will be the number of xy's, and CD the number derived from multiplying the coefficient of xy[31] by y, which is CF. Since, therefore, CD is the difference between AB and BD, this rule follows:

Rule

When a number of x's are equal to xy and the square of x, and x is known, we subtract this from its coefficient and divide the remainder by the coefficient of xy. What results is the value of y.[32]

[28] I.e., if $xy = ax + cx^2$, and if y is given, $x = \frac{y - a}{c}$.

[29] I.e., if $xy = ax + cx^2$, and if xy is given, $x = \sqrt{xy/c + (a/2c)^2} - a/2c$.

[30] The diagram in 1663 is mislabeled by having B instead of D in the upper right hand corner.

For example,

$$10x = x^2 + 3xy.$$

X equals 4. We subtract 4 from 10, leaving 6. Divide this by 3, the coefficient of xy and 2 is the result, the value of y.

If y is the known, we multiply this by the coefficient of xy, and subtract the product from the coefficient of x. The remainder is the value of x.[33]

For example,

$$10x = x^2 + 3xy.$$

Y equals 2. We multiply, therefore, 2, the value of y, by 3, the coefficient of xy. This produces 6 which I subtract from 10, the coefficient of x, leaving 4, the value of x.

If xy is the known, we multiply this by its coefficient and subtract the product from the square of one-half the coefficient of x. The root of the remainder added to or subtracted from the same half the coefficient of x, shows the value of x.[34]

For example,

$$10x = x^2 + 3xy.$$

XY equals 8. I multiply 8 by 3, the coefficient of xy, yielding 24. This we subtract from 25, the square of 5, which is half of 10, and 1 remains, the square root of which is 1. This added to or subtracted from 5 gives 6 and 4 as the values of x. Hence by dividing 8, the value of xy, by 6 or 4, either $1\frac{1}{3}$ or 2 is produced as the value of y.

Demonstration

9. If the square of x, xy, and y are equated in turn, there arise three other forms. For the first, let the square of x equal xy and a number of y's, and assume *AB* to be x, the square of which, *AC*, is equal to xy,

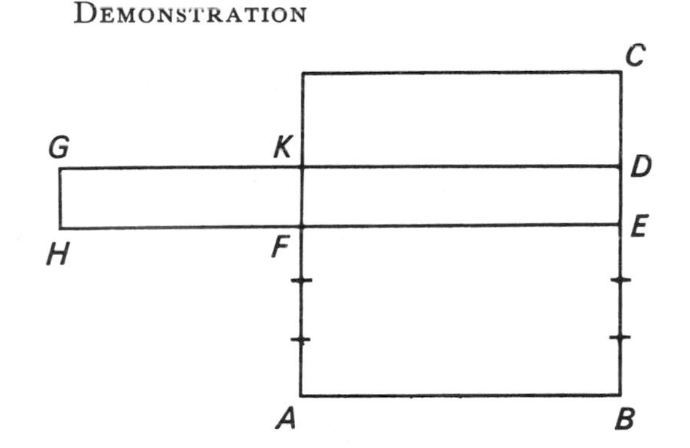

---

[31] The text has y.

[32] I.e., if $ax = bxy + x^2$, and if x is given, $y = \frac{a - x}{b}$.

[33] I.e., if $ax = bxy + x^2$, and if y is given, $x = a - by$.

[34] I.e., if $ax = bxy + x^2$, and if xy is given, $x = a/2 \pm \sqrt{(a/2)^2 - bxy}$.

which is AD (wherefore DE is y), plus y [or] DE, [whose] coefficient is FH. The surface GF will be equal by agreement to the surface CK. Therefore, by VI, 16[35] of the *Elements*,

$$AB:DE = HF:DC.$$

But AB is x, DE y, HF the coefficient of y, and CD the difference between x and the product of y and the coefficient of xy. Whence the rule is:

Rule

When the square of x is equal to the xy's and y's and x is the known, we multiply x by the coefficient of xy and to the product we add the coefficient of y and divide x^2 by the sum. The quotient is the value of y.[36]

For example,

$$x^2 = 6xy + 20y,$$

and x equals 12. I multiply 12 by 6, the coefficient of xy, making 72. To this I add 20, the coefficient of y, making 92. Then I divide 144,[37] the value of x^2, by this, and $1\frac{13}{23}$ is the result, the value of y itself.

If it is y that is known, we multiply this by its own coefficient and save the product [until later]. Then we multiply this same [value of y] by the coefficient of xy and add the square of one-half of this product to the first product and to the square root of the sum we add[38] the half that was squared, and the sum is the value of x.[39]

For example,

$$x^2 = 12y + 5xy,$$

and y equals 2. I multiply 2, the value of y, by 12, its coefficient, and 24 results. Then I multiply this same y, or 2, by 5, the coefficient of xy, making 10, half of which is 5. I square this, making 25, and add this to 24 — now being used — and get 49, the square root of which is 7.

[35] The text has VI, 15.

[36] I.e., if $x^2 = axy + by$, and if x is given, $y = \dfrac{x^2}{ax + b}$.

[37] 1570 has 244.

[38] 1545 has *adijciemus*; 1570 and 1643 have *abijciemus*.

[39] I.e., if $x^2 = axy + by$, and if y is given, $x = \sqrt{by + \left(\frac{ay}{2}\right)^2} + \frac{ay}{2}$.

To this I add the same half [as above], which is 5, producing 12, the value of x.

Where, however, it is xy that is known (and on the figure this is the surface EK), we multiply this by its own coefficient and cube one-third the product. We also multiply xy by the coefficient of y and square half the product. From this we subtract the part which we just cubed — that is the cube of one-third the first product that you saved — and the root of this remainder we add to and subtract from half the second product, and the sum of the cube roots of the sum and difference is the value of x.[40]

For example,

$$x^2 = 12y + 2xy,$$

and xy equals 24. I multiply 2 by 24, and get 48, one-third of which, which is 16, I cube, making 4096. I also multiply 24 by 12, and get 288, half of which, which is 144, I square, and this square is 20,736. From this I subtract 4096, leaving 16,640, the square root of which I add to and subtract from 144: that is $144 + \sqrt{16{,}640}$ and $144 - \sqrt{16{,}640}$, the sum of the cube roots of which is the value of x.

But if, having divided the number equally [i.e., into two parts], the square of half of it is not equal to or greater than the cube of one-third the first product, you will use throughout the remainder of the operation the rule for the case of the cube equal to the first power and constant, [the coefficient of the first power and the constant being the old coefficients] multiplied by the product [i.e., by the value of xy]. Thus in the example, [multiply] by 24, which number is the value of xy. Then the cube will be equal to the first power and the number, [the coefficient of] x being the product of xy and its coefficient and the constant being the product of xy and the coefficient of y. Thus it was said in the example that

$$x^2 = 2xy + 12y$$

and that xy equals 24. We must say that

$$x^3 = 48x + 288,$$

48 being the product of 2 and 24, and 288 that of 24 and 12.

[40] I.e., if $x^2 = axy + by$, and if xy is given,

$$x = \sqrt[3]{\frac{bxy}{2} + \sqrt{\left(\frac{bxy}{2}\right)^2 - \left(\frac{axy}{3}\right)^3}} + \sqrt[3]{\frac{bxy}{2} - \sqrt{\left(\frac{bxy}{2}\right)^2 - \left(\frac{axy}{3}\right)^3}}.$$

Let us assume that

$$x^2 = 2xy + 3y$$

and that xy equals 8. Multiply 8 by 2 and by 3 and thus produce 16 and 24. Therefore,

$$x^3 = 16x + 24$$

and x will equal $\sqrt{13} + 1$, according to the rule. Then I divide the value of xy, 8, by $\sqrt{13} + 1$, and there results $\sqrt{5\frac{7}{9}} - \frac{2}{3}$ as the value of y. However, the square of $\sqrt{13} + 1$ is $14 + \sqrt{52}$ and xy equals $\sqrt{75\frac{1}{9}} - \frac{2}{3}$, which is 8. Twice this is 16, and $3y$ is $\sqrt{52} - 2$ which, added to 16 or $2xy$, makes $14 + \sqrt{52}$,[41] the value of x^2.

Note 1. Note that in this rule, x is always the mean proportional between y and the coefficient of y plus the product of x and the coefficient of xy. Thus, in the example, $\sqrt{13} + 1$, which is x, is proportional between $\sqrt{5\frac{7}{9}} - \frac{2}{3}$, which is y, and $\sqrt{52} + 5$, which is composed of 3, the coefficient of y, plus the product of $\sqrt{13} + 1$, or x, and 2, the coefficient of xy.

Note 2. Note also that this rule derives from the case of the cube equal to the first power and constant, just as the next one comes from the case of the cube and constant equal to the first power and the last one from the case of the cube and first power equal to the constant.

Note 3. Note also that x is the same as was sought in the case of the cube equal to the first power and constant, but y is the number which comes from dividing the constant, whatever it may be, by this same x, for in the same case of the cube equal to the first power and constant, there is only one appropriate x, but there may be any number of y's.[42] In the given case, as has been said, x equals $\sqrt{13} + 1$, and we divided 8, the value of xy, [by it]. If, however, it is assumed that

$$x^3 = 16x + 24$$

x will always be $\sqrt{13} + 1$. But if we assumed xy to be 4, the coefficient of y would be 6, the coefficient of xy would be 4, and y would be $\sqrt{1\frac{4}{9}} - \frac{1}{3}$.[43]

[41] 1570 and 1663 have $\sqrt{25}$.

[42] *competit una sola res, sed infinitae quantitates.*

[43] The text has $\frac{2}{3}$.

Demonstration

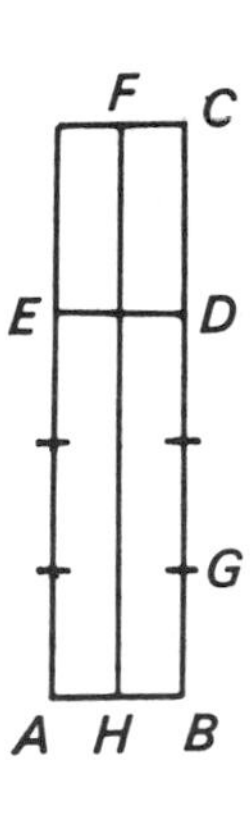

10. If xy equals the square of x plus a second unknown, we let AB be x, BC y, and the coefficient of x^2 the quotient of BG (which is equal to AB) in BD. The sum of the squares will be AD and the remainder, EC, will equal xy,[44] Let FC be the coefficient of y. Then FB will equal EC and [the ratio of] BC, or y, to AB, or x, will be [the same as the ratio of] DC (the difference between y and the product of x and the coefficient of x^2) to CF, the coefficient of y. From this, [it appears that] the remainder, EB, will be equal to the remainder, AF, and therefore that AB is the mean proportional between AH and BC divided by the number which is the quotient of BG in BD.

Note. Note, therefore, that throughout this rule x is a mean proportional between y divided by the coefficient of the square, and the difference between x and the coefficient of y.

Rule

The rule is that when xy is equal to y plus the square of x, and x is known, we square it and then multiply this [i.e., the square] by the coefficient of x^2, and we divide the product by the difference between x and the coefficient of y, and what results is y.[45]

For example,

$$xy = 3x^2 + 12y.$$

Let x equal 20, for example. I square 20 and get 400. I multiply 400 by 3, the coefficient of x^2, and get 1200. I divide 1200 by 8, the difference between x and the coefficient of y, which leaves 150, the value of y.

If it is y that is known, and not x, multiply this by its coefficient and divide the product by the coefficient of x^2. Subtract the result from the square of half the quotient of y over the coefficient of x^2. The square root of the remainder added to or subtracted from half of the foregoing quotient is the value of x.[46]

For example,

$$xy = 4x^2 + 3y.$$

[44] The text has "the coefficient of y."

[45] I.e., if $xy = ay + bx^2$, and if x is given, $y = \frac{bx^2}{x - a}$.

[46] I.e., if $xy = ay + bx^2$, and if y is given, $x = y/2b \pm \sqrt{(y/2b)^2 - ay/b}$.

Y equals 50. Multiply 50 by 3, the coefficient of y, and 150 results. Divide 150 by 4, the coefficient of x^2, making $37\frac{1}{2}$. Then divide 50 by 4 — that is, y by the coefficient of x^2 — and $12\frac{1}{2}$ results. Square half of this, which is $6\frac{1}{4}$, making $39\frac{1}{16}$, from which subtract $37\frac{1}{2}$, leaving $1\frac{9}{16}$, the square root of which is $1\frac{1}{4}$ which, added to or subtracted from $6\frac{1}{4}$, shows the value of x to be $7\frac{1}{2}$ or 5.

If, however, it is the product, or xy, that is known, we multiply xy by the coefficient of y and divide the product by the coefficient of x^2. The result is a number which with x^3 is equal to as many x's as is the number resulting from the division of xy by the coefficient of x^2.[47]

For example,

$$xy = 4x^2 + 6y,$$

and xy equals 1500. Accordingly, we multiply 1500 by 6, making 9000 which, divided by 4, the coefficient of x^2, makes 2250, a number which, with x^3 added to it, should equal $375x$, 375 being the number which results from dividing 1500, or xy,[48] by 4, the coefficient of x^2. According to the proper rule, then, x equals 15 or $\sqrt{206\frac{1}{4}} - 7\frac{1}{2}$,[49] and either of these numbers may be the solution for x. In this case, when xy, which is 1500, is equal to $4x^2 + 6y$, and the value of y is wanted, I divide 1500, which is the value of xy, by either solution for x.

Demonstration

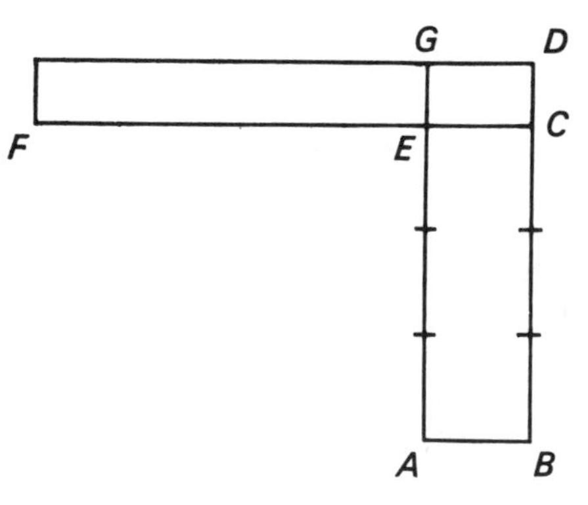

11. If y, or CD, multiplied by CF, its coefficient, is equal to the square of x, or AB, and to DE, or xy, this having been reduced to a coefficient of 1, let the common surface DE be subtracted. Then the surface GF will be equal to AC. Hence the square of AB, according to VI, 1 of the *Elements*, will be equal to the surface produced by multiplying EG by such a portion of EF as AB is of BC. Therefore, according

[47] I.e., if $xy = ay + bx^2$, and if xy is given, $x^3 + \frac{axy}{b} = \frac{xy}{b}x$.

[48] The text says 1500 is the coefficient of xy.

[49] The text has 10 or $\sqrt{300} - 5$.

to VI, 16 of the *Elements*, *AB* is a mean between *DC* and that part of *EF*. Whence the rule:

Rule

When a second unknown is equal to xy and the square of x, and x is the known, we square this, then multiply the product by the coefficient of x^2, and then divide the last product by the coefficient of y, having subtracted x, and the result is the value of y.[50]

For example,

$$12y = xy + 3x^2.$$

X equals 4. I square 4 and get 16. I multiply 16 by 3, the coefficient of x^2, making 48. I divide 48 by 12, the coefficient of y, minus x, or 4 — this is equivalent to dividing by 8 — and 6, or the value of y, results.

If y is the known, multiply this by its coefficient and divide the product by the coefficient of x^2. To the quotient add the square of half of that which results from dividing y by the coefficient of x^2 and the square root of the whole, minus this same half, is the value of x.[51]

For example,

$$12y = xy + 3x^2.$$

Y equals 6. I multiply 12 by 6, making 72. This I divide by 3, the coefficient of x^2, making 24. [52]Then I divide 6, or y, by 3, the coefficient of x^2, making 2,[52] half of which is 1. This squared is likewise 1. I add this to 24, making 25, the square root of which is 5, and subtracting 1, half of 2, leaves 4 as the value of x.

If it is xy that is the known, we multiply this by the coefficient of y and divide the product by the coefficient of x^2 and what results is a number equal to a cube and first power, the coefficient of the latter of which is the quotient of xy divided by the coefficient of x^2. Hence the value of x is the solution sought for and, dividing xy by this, the value of y results.[53]

[50] I.e., if $ax^2 + xy = by$, and if x is given, $y = \frac{ax^2}{b - x}$.

[51] I.e., if $ax^2 + xy = by$, and if y is given, $x = \sqrt{\frac{by}{a} + \left(\frac{y}{2a}\right)^2} - \frac{y}{2a}$.

[52] 1663 includes this passage twice with, however, 24 in place of 2 the first time.

[53] I.e., if $ax^2 + xy = by$, and if xy is given, $\frac{bxy}{a} = x^3 + \left(\frac{xy}{a}\right)x$.

For example,

$$12y = xy + 3x^2.$$

XY equals 24. I multiply 24 by 12, making 288. This I divide by 3, [54]getting 96. Then I divide 24 by this same 3,[54] the coefficient of x^2, and get 8. Hence

$$x^3 + 8x = 96.$$

Then, according to its own proper rule, x will equal 4. Therefore 4 is the value of x. Divide 24, or xy, by this and 6 results as the value of y.

Note. You know that every case or rule from among the foregoing has all the properties inherent in the rule [when applied] in particular instances. Hence we use one method in one case and another in another, depending on what is known. As an example: The 10th rule has five properties: First, that the ratio of y to x is that of y minus the product of x and the coefficient of x^2 to the coefficient of y.[55] Second, that x is a proportional[56] between y divided by the coefficient of x^2, and the difference between x and the coefficient of y.[57] Third, that the result of squaring x and multiplying this by the coefficient of x^2, having derived x^2, is the same as the product of y and the difference between x and the coefficient of y.[58] The fourth and fifth are that there are two other methods of proceeding under this rule to the discovery of x, examples of which we give in the problems.

Problem I

Find two numbers the sum of whose squares is 100 and the product of one of which by the other is equal to twice their sum.[59]

Let x be one and y the other. Hence

$$xy = 2x + 2y.$$

[54] 1663 omits these words.

[55] *Prima, quod proportio quantitatis ad rem, est ut ducta re in numerum quadratorum, et detracta quantitate, ad numerum quantitatum.* I.e., if $xy = ax^2 + by, \frac{y}{x} = \frac{y - ax}{b}$.

[56] 1545 has *proportionalis*; 1570 and 1663 have *media proportione*. This difference in language occurs frequently throughout the book and will not be noted hereafter.

[57] I.e., if $xy = ax^2 + by, \frac{y}{a} : x = x : (x - b)$.

[58] I.e., if $xy = ax^2 + by$, $ax^2 = y(x - b)$.

[59] The Latin is somewhat obscure, but the context makes it clear that it should be translated as given: *Invenias duos numeros, quorum quadrata iuncta, sint 100, et productus unius in alterum duplum sit aggregato eorum.*

According to the fourth rule, therefore, the ratio of the remainder of x [i.e., $x - 2$] to 2 is as 2 to the remainder of y [i.e., $y - 2$]. Hence there will be three quantities in proportion: the remainder of x, 2, and the remainder of y. X, however, consists of its remainder plus 2, and y of its remainder plus 2. Therefore x is the sum of the first and second of the three proportional quantities, and y of the second and third. According to the rules in the chapter on three proportional quantities, then, [60] the square of the sum of the first and second plus [60] the square of the sum of the second and third plus the square of the second is equal to the square of the sum of all three. But, by agreement, the square of the sum of the first and second and the square of the sum of the second and third is 100, and the square of the second is 4, since the second proportional quantity is 2. Therefore the square of the sum of all three quantities is 104 and the sum of these three quantities is $\sqrt{104}$. Since the second is 2, the others — namely, the first and third — will be $\sqrt{104} - 2$. Divide, then, $\sqrt{104} - 2$ into two parts, the product of which is 4, the square of 2, and these will be $\sqrt{26} - 1 + \sqrt{23 - \sqrt{104}}$, and $\sqrt{26} - 1 - \sqrt{23 - \sqrt{104}}$. Since x consists of the first and second proportionals [61] and y of the second and third, [61] we add 2 — that is, the second quantity — to both and this makes x equal to $\sqrt{26} + 1 + \sqrt{23 - \sqrt{104}}$ and y equal to $\sqrt{26} + 1 - \sqrt{23 - \sqrt{104}}$. The sum of the squares of these is exactly 100, and their product is $\sqrt{416} + 4$, which is twice their sum. By the common way of proceeding you would arrive at

$$\sqrt{50 + \sqrt{2068 - \sqrt{26{,}624}}} \quad \text{and} \quad \sqrt{50 - \sqrt{2068 - \sqrt{26{,}624}}}.$$

It is evident, however, that these are more confusing than, though they are equivalent to, the ones given above.

PROBLEM II

Find two numbers such that the sum of their squares is 100 and the square of the greater is equal to the greater times four times the less plus eight times the greater.

Let x be the greater and y the less. Therefore,

$$x^2 = 4xy + 8x,$$

[60] 1663 omits this passage.

[61] 1663 omits this passage.

wherefore, according to the sixth rule, we subtract 8 from x, which makes the remainder $x - 8$. This, divided by 4, will be $\frac{1}{4}x - 2$, and this is y. Therefore the squares of x and of $\frac{1}{4}x - 2$ are 100.[62] Hence

$$1\tfrac{1}{16}x^2 + 4 - x = 100$$

and

$$x^2 = \tfrac{16}{17}x + 90\tfrac{6}{17},$$

wherefore x equals $\sqrt{90\frac{166}{289}} + \frac{8}{17}$, and y is one-fourth of this minus 2, namely $\sqrt{5\frac{191}{289}} - 1\frac{15}{17}$.

Problem III

Find two numbers the sum of whose squares is 100, and the product of one of which times the other is equal to three times the square of the smaller plus six times this same smaller one.

Let x represent the smaller number and y the greater. Therefore

$$xy = 3x^2 + 6x.$$

Hence, according to the seventh rule,

$$y = 3x + 6.$$

Therefore,

$$x^2 + (3x + 6)^2 = 100,$$

and

$$10x^2 + 36x + 36 = 100,$$

wherefore

$$x^2 + 3\tfrac{3}{5}x = 6\tfrac{2}{5}.$$

X is, therefore, $\sqrt{9\frac{16}{25}} - 1\frac{4}{5}$, and y is three times this plus 6, that is $\sqrt{86\frac{19}{25}} + \frac{3}{5}$.[63]

Problem IIII

Divide 20 into three proportional parts,[64] the square of the mean of which is equal to twice the product of the mean and least plus four times the least.

[62] 1570 and 1663 have 110.

[63] 1570 and 1663 have $\sqrt{86\frac{21}{25}} + \frac{3}{5}$.

[64] 1545 has *tres partes proportionales*; 1570 and 1663 have *tres partes in continua proportione*. This variance in language occurs frequently throughout the book and will not be further noted.

Let x be the mean and y the least. Then

$$x^2 = 2xy + 4y.$$

Wherefore, from the first note to the ninth rule, x is a mean proportional between y and the sum of 4, the coefficient of y, and the product of x and the coefficient of xy, namely 2. The third quantity, therefore, is $2x + 4$. Since, then, the third quantity is $2x + 4$, and the second quantity is x, and these plus the first are 10, the first will be $6 - 3x$. Therefore, by multiplying the first and third, the square of the second is produced. Accordingly,

$$x^2 = 24 - 6x^2,$$

wherefore $7x^2$ equals 24, and x is $\sqrt{3\frac{3}{7}}$,[65] and this is the mean quantity, twice which, plus 4, is the third, namely $4 + \sqrt{13\frac{5}{7}}$. Hence, by subtracting the sum of the second and third from 10, the first is left as $6 - \sqrt{30\frac{6}{7}}$. These, moreover, are proportional quantities and the square of the second is equal to twice the product of the second and first plus four times the first, as was proposed.

[65] 1570 and 1663 have $\sqrt{3\frac{5}{7}}$.

CHAPTER XI

On the Cube and First Power Equal to the Number

Scipio Ferro of Bologna well-nigh thirty years ago discovered this rule and handed it on to Antonio Maria Fior of Venice, whose contest with Niccolò Tartaglia of Brescia gave Niccolò occasion to discover it. He [Tartaglia] gave it to me in reponse to my entreaties, though withholding the demonstration. Armed with this assistance, I sought out its demonstration in [various] forms. This was very difficult. My version of it follows.

DEMONSTRATION

For example, let GH^3 plus six times its side GH equal 20, and let AE and CL be two cubes the difference between which is 20 and such that the product of AC, the side [of one], and CK, the side [of the other], is 2, namely one-third the coefficient of x. Marking off BC equal to CK, I say that, if this is done, the remaining line AB is equal to GH and is, therefore, the value of x, for GH has already been given as [equal to x].

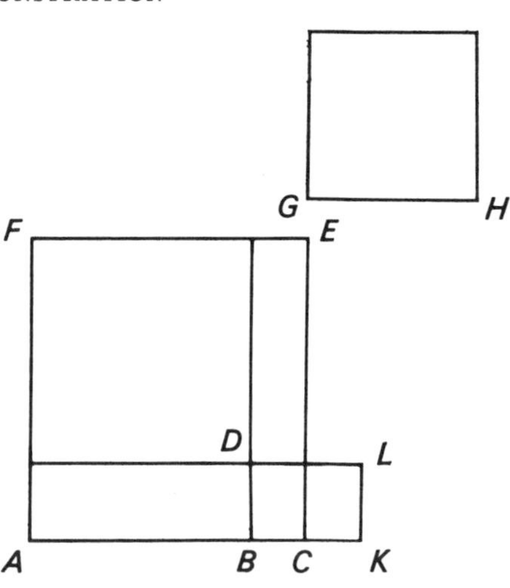

In accordance with the first proposition of the sixth chapter of this book, I complete the bodies DA, DC, DE, and DF; and as DC represents BC^3, so DF represents AB^3, DA represents $3(BC \times AB^2)$ and

DE represents $3(AB \times BC^2)$.[1] Since, therefore, $AC \times CK$ equals 2, $AC \times 3CK$ will equal 6, the coefficient of x; therefore $AB \times 3(AC \times CK)$ makes $6x$ or $6AB$,[2] wherefore three times the product of AB, BC, and AC is $6AB$. Now the difference between AC^3 and CK^3 — manifesting itself as BC^3, which is equal to this by supposition — is 20, and from the first proposition of the sixth chapter is the sum of the bodies DA, DE, and DF. Therefore these three bodies equal 20.

Now assume that BC is negative:

$$AB^3 = AC^3 + 3(AC \times CB^2) + (-BC^3) + 3(-BC \times AC^2),$$

by that demonstration. The difference between $3(BC \times AC^2)$ and $3(AC \times BC^2)$, however, is [three times] the product of AB, BC, and

[1] 1570 and 1663 vary considerably from here on to the end of the demonstration. They read:

We will have, therefore, four propositions, two of which have already been mentioned — namely, that $AC \times CK$ or CB is 2 and that the difference between AC^3 and CB^3 is 20. The third can be deduced from these and is that, since the product of $3AB \times BC \times AC$ is equal to the sum of [the text has, "the difference between"] DE and DA and that $3AB \times AC \times BC$ is $6AB$ for, from the first proposition, the product of AC and CB is 2, therefore three times this is 6, and this product times AB is $6AB$. This, however, is the sum of [the text has, "the difference between"] DE and DA. The fourth [proposition], which derives from the second and third corollaries in the sixth chapter, [is] that DF [i.e., AB^3] is the difference between $AC^3 + 3(AC \times CB^2)$ and $CB^3 + 3(CB \times AC^2)$. Let, therefore, α be AC^3, β BC^3 [the text has ABC^3], γ $3CB \times AC^2$, δ $3AC \times CB^2$, ϵ the difference between α and β, ζ the difference between γ and δ, and η [1570 has β; 1663's character is illegible] the difference between $\alpha + \delta$ and $\beta + \gamma$. Therefore [there is what appears to be a superfluous *cum* inserted at this point] ϵ is composed of $\zeta + \eta$ [1570 again has a β and again 1663's character is illegible], as can readily be demonstrated numerically and by example as shown in the margin. But ϵ is 20, from the second assumption, ζ is $6AB$, and η [1570's character is illegible and 1663 has θ] is AB^3. Therefore, $AB^3 + 6AB$ — that is, plus $6x$, for AB is the root of its cube — is equal to 20. Therefore, since $GH^3 + 6GH$ [the text has $BH^3 + 6BH$ here and the next place it occurs] is equal to 20, GH^3+6GH will be equal to AB^3+6AB. Hence AB is x and this is the difference between two sides the product of which is 2 and the cubes of which differ by 20, which was to be demonstrated. From this we construct the rule.

	24	1		25	
20			13		7
	4	14		18	

As it is obvious from the preceding the text of 1570 and 1663 is quite corrupt. These items are especially bothersome:

(1) the *differentiae DE et DA* which occurs twice. To make sense, we have either to assume, as I have here, that *differentia* is an error for *aggregatum* or that, in the earlier definitions of DA as *triplus CB in quadratum AB* and of DE as *triplus, AB in quadratum BC*, AB should be replaced by AC. Either is consistent with the later development of the argument.

(2) the *cubus abc* which I read as a typographical error for *cubus bc*.

(3) the confusion between eta, beta, and theta at various points.

(4) the misprinting of BH for GH at four places.

[2] *fiunt 6 res AB, seu sexcuplum AB.*

AC. Therefore, since this, as was demonstrated, is equal to $6AB$, add $6AB$ to the product of $3(AC \times BC^2)$, making $3(BC \times AC^2)$. But since BC is negative, it is now clear that $3(BC \times AC^2)$ is negative and the remainder which is equal to it is positive. Therefore,

$$3(CB \times AB^2) + 3(AC \times BC^2) + 6AB = 0.$$[3]

It will be seen, therefore, that as much as is the difference between AC^3 and BC^3, so much is the sum of

$$AC^3 + 3(AC \times CB^2) + 3(-CB \times AC^2) + (-BC^3) + 6AB.$$

This, therefore, is 20 and, since the difference between AC^3 and BC^3 is 20, then, by the second proposition of the sixth chapter, assuming BC to be negative,

$$AB^3 = AC^3 + 3(AC \times BC^2) + (-BC^3) + 3(-BC \times AC^2).$$

Therefore since we now agree that

$$AB^3+6AB\text{[4]} = AC^3+3(AC\times BC^2)+3(-BC\times AC^{2}\text{[5]})+(-BC^3)+6AB,$$

which equals 20, as has been proved, they [i.e., $AB^3 + 6AB$] will equal 20. Since, therefore,

$$AB^3 + 6AB = 20,$$

and since

$$GH^3 + 6GH = 20,$$

it will be seen at once and from what is said in I, 35 and XI, 31 of the *Elements* that GH will equal AB. Therefore GH is the difference between AC and CB. AC and CB, or AC and CK, the coefficients, however, are lines containing a surface equal to one-third the coefficient of x and their cubes differ by the constant of the equation. Whence we have the rule:

Rule

Cube one-third the coefficient of x; add to it the square of one-half the constant of the equation; and take the square root of the whole. You will duplicate[6] this, and to one of the two you add one-half the number you have already squared and from the other you subtract one-half the same. You will then have a *binomium* and its *apotome*. Then,

[3] *faciunt nihil.*

[4] The text has a spare *cum* at this point, as though something more were to be added.

[5] 1545 has AB^2.

[6] 1545 has *seminabis*; 1570 and 1663 have *servabis*. The former, corrected to read *geminabis* in accord with later passages, is followed here.

subtracting the cube root of the *apotome* from the cube root of the *binomium*, the remainder [or] that which is left is the value of x.[7]

For example,

$$x^3 + 6x = 20.$$

Cube 2, one-third of 6, making 8; square 10, one-half the constant; 100 results. Add 100 and 8, making 108, the square root of which is $\sqrt{108}$. This you will duplicate: to one add 10, one-half the constant, and from the other subtract the same. Thus you will obtain the *binomium* $\sqrt{108} + 10$ and its *apotome* $\sqrt{108} - 10$. Take the cube roots of these. Subtract [the cube root of the] *apotome* from that of the *binomium* and you will have the value of x:

$$\sqrt[3]{\sqrt{108} + 10}\,^{8} - \sqrt[3]{\sqrt{108} - 10}$$

Again,

$$x^3 + 3x = 10.$$

Cube 1, one-third of 3, and 1 results; square 5, one-half of 10, and 25 results; add 25 and 1, making 26; add 5 to and subtract it from the square root of this. You will thus form the *binomium* $\sqrt{26} + 5$ and its *apotome* $\sqrt{26} - 5$; whence x equals $\sqrt[3]{\sqrt{26} + 5} - \sqrt[3]{\sqrt{26} - 5}$. Here you have the proof:

	$\sqrt[3]{\sqrt{26} + 5}$	$-\sqrt[3]{\sqrt{26} - 5}$[9]
The cubes of the parts: (As is evident, the sum of these is 10.)	$(\sqrt{26} + 5)$	$-(\sqrt{26} - 5)$
The squares of the parts:	$\sqrt[3]{51 + \sqrt{2600}}$[10]	$\sqrt[3]{51 - \sqrt{2600}}$
Three times the squares of the parts:	$\sqrt[3]{1377 + \sqrt{1,895,400}}$[11]	$\sqrt[3]{1377 - \sqrt{1,895,400}}$[12]
The parts themselves:	$-\sqrt[3]{\sqrt{26} - 5}$	$+\sqrt[3]{\sqrt{26} + 5}$
The products of the parts and three times their squares:	$+\sqrt[3]{\sqrt{49,299,354} + 6885 - \sqrt{47,385,000} - 7020}$	$-\sqrt[3]{\sqrt{49,299,354} - 6885 - \sqrt{47,385,000} + 7020}$

[7] I.e., if $x^3 + ax = N$, $x = \sqrt[3]{\sqrt{(a/3)^3 + (N/2)^2} + N/2} - \sqrt[3]{\sqrt{(a/3)^3 + (N/2)^2} - N/2}$.

[8] 1570 and 1663 have ℞ *b: cub:* ℞ 108 *p:* 10, the *b:* being a misprint for *v:*.

[9] 1570 and 1663 have $\sqrt[3]{\sqrt{27} - 5}$.

[10] 1570 and 1663 have $\sqrt[3]{51 + \sqrt{2900}}$.

[11] 1570 and 1663 have $\sqrt[3]{1277 - \sqrt{1,895,400}}$.

[12] 1570 and 1663 have $\sqrt[3]{1377 - \sqrt{1,865,400}}$.

Moreover, the cube roots contain four terms which can be reduced to two, for when 6885 is subtracted from 7020, the remainder is 135, and likewise when $\sqrt{47{,}385{,}000}$ is subtracted from $\sqrt{49{,}299{,}354}$ there is left $\sqrt{18{,}954}$.[13] Therefore these products are

$$\sqrt[3]{\sqrt{18{,}954} - 135} - \sqrt[3]{\sqrt{18{,}954} + 135}.$$

The whole cube, then, from the demonstration in the third book is $10 + \sqrt[3]{\sqrt{18{,}954} - 135} - \sqrt[3]{\sqrt{18{,}954} + 135}$, and three times the root, or $3x$, equals $\sqrt[3]{\sqrt{18{,}954} + 135} - \sqrt[3]{\sqrt{18{,}954} - 135}$. And, finally, having added all together, since the universal cube roots cancel each other, the whole becomes

$$x^3 + 3x = \text{exactly } 10.$$

A third example:

$$x^3 + 6x = 2.$$

Raise 2, one-third the coefficient of x, to the cube and 8 is the result; square 1, half of 2, making 1; add 8 to 1, and 9 is produced, the square root of which is 3. Now duplicate 3 and to one add 1, half the constant, thus making 4, and from the other subtract half the constant, thus making 2. Then subtract the cube root of the less from the cube root of the greater and you have $\sqrt[3]{4} - \sqrt[3]{2}$[14] as the value of x.

Remember what we said in the chapter in the third book on extracting cube roots whenever these universal cube roots are equivalent to a whole number or a fraction. Thus in the first example

$$\sqrt[3]{\sqrt{108} + 10} - \sqrt[3]{\sqrt{108} - 10}$$

is 2, as is indicated by the rule there given and as is perfectly clear if it is tried out.[15]

[13] 1663 has $\sqrt{18{,}854}$.

[14] 1663 has $\sqrt[3]{1}$.

[15] 1570 and 1663 add:

It is easy to understand both in this and in the following chapters that, having the solution and the coefficient of x, we obtain the constant of the equation by multiplying the solution by the coefficient of x and adding to this product the cube of the same [i.e., the cube of the value of x], for the sum is the constant of the equation. Thus, given $x^3 + 3x$ and the value of x as 2, I say that you should multiply 2 by 3, making 6, and add this to 8, the cube of 2, making 14, the constant of the equation. Similarly, given x^3 plus a certain number of x's equal to 20 and the value of x [the text seems to say "the value of x^3"] as $\sqrt{8} - 2$, for example, we can derive the coefficient of x by raising $\sqrt{8} - 2$ to the cube, making $\sqrt{3200} - 56$, subtracting this from the constant, which is 20, leaving $76 - \sqrt{3200}$, and dividing this by $\sqrt{8} - 2$, the solution, making $\sqrt{648} - 2$, the coefficient of x.

You also know that there may be a solution common to all types of cases as, for instance, of the cube and constant equal to the first power. Thus if

$$x^3 + 12 = 34x,$$

and the value of x is $3 + \sqrt{7}$ or $3 - \sqrt{7}$, and I wish x^3 plus a number of x's to equal 12 with the latter solution, the number of x's will be, according to the preceding rule, $\sqrt{1008} + 2$. By following the procedure of this chapter and taking one-third the coefficient of x, which is $\sqrt{112} + \frac{2}{3}$, and raising it to the cube, making $\sqrt{1,438,577\frac{7}{9}} + 224\frac{8}{27}$, and adding 36, the square of one-half of 12, the constant of the equation, you will have $\sqrt{1,438,577\frac{7}{9}} + 260\frac{8}{27}$. Add 6 to or subtract it from [the square root of] this and take the cube root [of the whole] and you will have the value of x, viz., $\sqrt[3]{\sqrt{\sqrt{1,438,577\frac{7}{9}} + 260\frac{8}{27}} + 6} - \sqrt[3]{\sqrt{\sqrt{1,438,577\frac{7}{9}} + 260\frac{8}{27}} - 6}$.

Note: In lieu of $\sqrt{1,438,577\frac{7}{9}}$ at the four places at which it appears above, 1570 and 1663 have these figures:

1570:	$\sqrt{1,905,552}$	$\sqrt{1,905,552}$	$\sqrt{190,555}$	$\sqrt{190,555}$
1663:	$\sqrt{1,905,552}$	$\sqrt{1,905,552}$	$\sqrt{1,905,552}$	$\sqrt{190,555}$

1570 also has in the first term *cui* instead of *cub* and in the second *ut* instead of *cub* to indicate that these are universal cube roots. It also has a superfluous R̸ in front of the $260\frac{8}{27}$ the first time it occurs. 1663 omits any indication that the second term is a *cube* root.

CHAPTER XII

On the Cube Equal to the First Power and Number

Demonstration

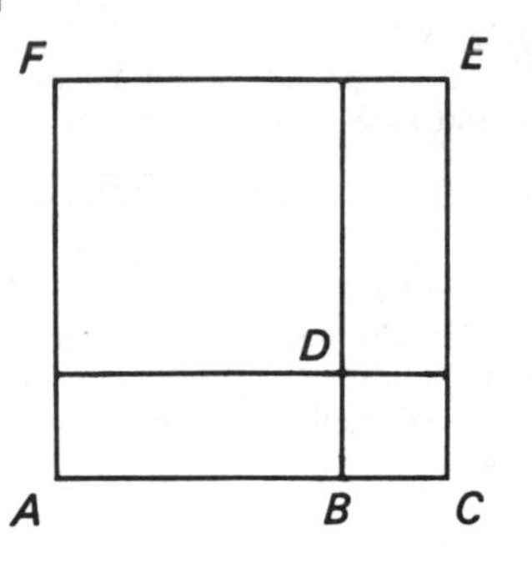

Let the cube be equal to the first power and constant and let DC and DF[1] be two cubes the product of the sides of which, AB and BC, is equal to one-third the coefficient of x, and let the sum of these cubes be equal to the constant. I say that AC is the value of x. Now since $AB \times BC$ equals one-third the coefficient of x, $3(AB \times BC)$ will equal the coefficient of x, and the product of AC and $3(AB \times BC)$ is the whole first power, AC having been assumed to be x. But $AC \times 3(AB \times BC)$ makes six bodies, three of which are $AB \times BC^2$ and the other three, $BC \times AB^2$. Therefore these six bodies are equal to the whole first power, and these [six bodies] plus the cubes DC and DF constitute the cube AE, according to the first proposition of Chapter VI. The cubes DC and DF are also equal to the given number. Therefore the cube AE is equal to the given first power and number, which was to be proved.

It remains to be shown that $3AC(AB \times BC)$ is equal to the six bodies. This is clear enough if I prove that $AB(BC \times AC)$ equals the two bodies $AB \times BC^2$ and $BC \times AB^2$, for the product of AC and $(AB \times BC)$ is equal to the product of AB and the surface BE — since all sides are equal to all sides[2] — but this [i.e., $AB \times BE$] is equal to the product of AB and $(CD + DE)$; the product $AB \times DE$ is equal to the product $CB \times AB^2$, since all sides are equal to all sides; and therefore $AC(AB \times BC)$ is equal to $AB \times BC^2$ plus $BC \times AB^2$, as was proposed.

[1] 1570 and 1663 have *DE*.

[2] *latera enim omnia omnibus sunt aequalia.*

Rule

The rule, therefore, is: When the cube of one-third the coefficient of x is not greater than the square of one-half the constant of the equation, subtract the former from the latter and add the square root of the remainder to one-half the constant of the equation and, again, subtract it from the same half, and you will have, as was said, a *binomium* and its *apotome*, the sum of the cube roots of which constitutes the value of x.[3]

For example,

$$x^3 = 6x + 40.$$

Raise 2, one-third the coefficient of x, to the cube, which makes 8; subtract this from 400, the square of 20, one-half the constant, making 392; the square root of this added to 20 makes 20[4] + $\sqrt{392}$, and subtracted from 20 makes $20 - \sqrt{392}$; and the sum of the cube roots of these, $\sqrt[3]{20 + \sqrt{392}} + \sqrt[3]{20 - \sqrt{392}}$, is the value of x.

Again,

$$x^3 = 6x + 6.$$

Cube one-third the coefficient of x, which is 2, making 8; subtract this from 9, the square of one-half of 6, the constant of the equation, leaving 1; the square root of this is 1; this added to and subtracted from 3, one-half the constant, makes the parts 4 and 2, the sum of the cube roots of which gives us $\sqrt[3]{4} + \sqrt[3]{2}$ for the value of x.

When the cube of one-third the coefficient of x is greater than the square of one-half the constant of the equation, which happens whenever the constant is less than three-fourths of this cube or when two-thirds of the coefficient of x muliplied by the square root of one-third the same number is greater than the constant of the equation, [5]then the solution of this can be found by the aliza problem which is discussed in the book of geometrical problems. But if you wish to avoid such a difficulty you may, for the most part, be satisfied by Chapter XXV of this work.[5]

[3] I.e., if $x^3 = ax + N$ and $(a/3)^3 \lesseqqgtr (N/2)^2$, then

$$x = \sqrt[3]{N/2 + \sqrt{(N/2)^2 - (a/3)^3}} + \sqrt[3]{N/2 - \sqrt{(N/2)^2 - (a/3)^3}}.$$

[4] 1570 and 1663 omit "makes 20."

[5] In lieu of this material, 1570 and 1663 have "then consult the Aliza book appended to this work."

CHAPTER XIII

On the Cube and Number Equal to the First Power

Demonstration

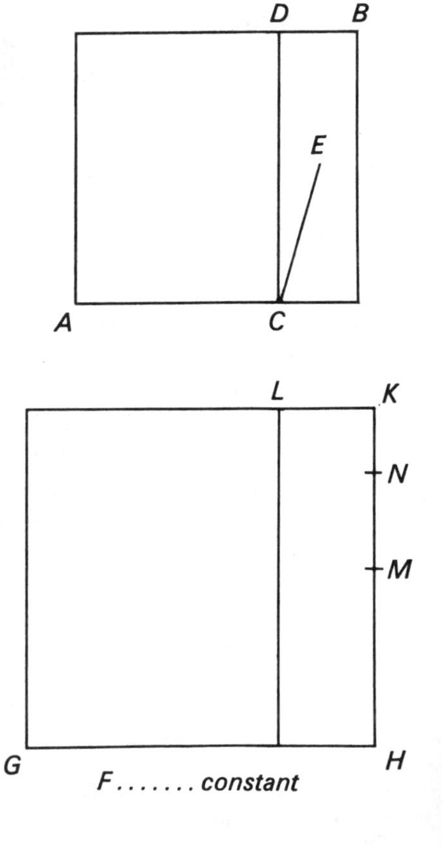

This rule derives from the preceding. Let the cube of GH be equal to the first power [with a coefficient of] AB (here shown as a square surface) plus F, the constant. And let the base of GH^3 be the square GK, one-fourth of which is HL, and let the remainder be equal to the surface AD. Let what in Greek is called the tetragonic root of the remainder, CB,[1] be CE, and let MK be one-half of HK. From this let MN be struck off equal to CE. I say that either HN^3 or NK^3 plus the number F is equal to AB x's, the coefficient of x and the constant of the equation remaining the same.

First let us show this with respect to HN. It is clear that HN^3 is made up of its side HN times HN^2. The square AB, however (since GL equals AD, and GL equals $3HM^2$), is equal to

[1] The text has *CD*.

$3HM^2 + MN^2$. These, however, according to II, 4[2] of the *Elements*, are greater than HN^2 by $2(HM \times NK)$[3] and, therefore, by the product $HK \times NK$, since HK is equal to $2HM$. HN^3, accordingly, is made up of its side HN times the surface AB minus the product $HK \times KN$.[4] And since GH^3[5] was composed of x, or its side HK, times HK^2 or of x times AB plus the constant F, therefore, by common consent, the number F is equal to the product of HK and the difference between the squares AB and GK. But the difference between GK and AB is equal to the difference between HL and CB, since AD is equal to GL. Furthermore, the difference between HL and CB is [the same] as that between HM^2 and MN^2. Therefore HK times the difference between HM^2 and MN^2 is the number F. Now, according to II, 4 of the *Elements*,[6] the difference between HM^2, or MK^2, and MN^2 is $2(MN \times NK) + NK^2$ and, therefore, is $(MN + MK)NK$ and, therefore, is $HN \times NK$. Hence the product of HK and $(HN \times NK)$ is F, the number. Therefore add F, the number, to HN^3 and, on the other side, the product $HK \times HN \times KN$ to the product of HN and the surface AB minus the product of HK and KN.[7] And this makes HN^3 plus the number F[8] equal to $HN \times AB$ or to AB x's, which was to be proved.

Likewise, since KH times the difference between GK and AB, which difference is $HN \times KN$, produces F, the difference between AB and KN^2 (since AB is equal to $HM^2 + MK^2 + MN^2 + (KM \times MH)$) is equal to the difference between $2(KH \times HN)$ and NH^2, having added the rectangle $HN \times NK$ to the latter.[9] But the product of HN and NK plus HN^2 is equal to the product of KH and HN, according to II, 3 of the *Elements*. Therefore the square AB is greater than NK^2 by the product of KH and HN. Since, then, the number F contains $NK \times (KH \times HN)$[10] and KN^3 contains $KN \times KN^2$, $KN^3 + F$, the number, or plus the product of KN and the rectangle $KH \times HN$, will be equal

[2] 1570 and 1663 omit this figure.

[3] 1570 and 1663 have $2(HM \times NH)$.

[4] I.e., $HN^3 = HN[AB - (HK \times KN)]$.

[5] The text has GK^3.

[6] 1570 and 1663 omit the citation, using *per eandem* instead.

[7] I.e., $F + HN^3 = (HK \times KN \times HN) + HN[AB - (HK \times KN)]$.

[8] 1545 omits the F.

[9] I.e., $AB - KN^2 = 2(KH \times HN) - [HN^2 + (HN \times NK)]$.

[10] 1570 and 1663 have $KH \times (KH \times HN)$.

to the product of AB and KN. Therefore KN^3 plus this same number F is equal to [the first power, KN, times] AB, the coefficient of the x's.

[11] From this demonstration it is evident that the solution for the cube equal to the first power and number is equal to the sum of the solutions for the cube plus the same number equal to the first power with the same coefficient. Thus, if

$$x^3 = 10x + 12$$

and the solution is $\sqrt{7} + 1$, the solutions for

$$x^3 + 12 = 10x$$

which are 2 and $\sqrt{7} - 1$, when added together make $\sqrt{7} + 1$.[11]

RULE

The rule, therefore, is: When the cube and constant are equal to the first power, find the solution for the cube equal to the same number of y's and the same constant; take three times the square of one-half of this and subtract it from the coefficient of the first power; and the square root of the remainder added to or subtracted from one-half the solution for the cube equal to y plus the constant gives the solution for the cube and constant equal to x.[12]

Example:

$$x^3 + 3 = 8x.$$

Solving

$$y^3 = 8y + 3$$

according to the preceding rule,[13] I obtain 3. The square of one-half of this is $2\frac{1}{4}$, which multiplied by 3 is $6\frac{3}{4}$. Subtracting this from 8, the coefficient of x, leaves $1\frac{1}{4}$, the square root of which added to or subtracted from $1\frac{1}{2}$, which is one-half the solution for the cube equal to the first power and constant, gives both solutions which were being sought. One is $1\frac{1}{2} + \sqrt{1\frac{1}{4}}$,[14] the other $1\frac{1}{2} - \sqrt{1\frac{1}{4}}$.

[11] 1570 and 1663 omit this passage.

[12] I.e., if $x^3 + N = ax$, and $y^3 = ay + N$, $x = y/2 \pm \sqrt{a - 3(y/2)^2}$.

[13] On the contrary, this problem is not workable by the preceding rule, which leads to an irreducible quantity.

[14] 1570 has $1\frac{1}{2} + \sqrt{1\frac{11}{4}}$.

Demonstration[15]

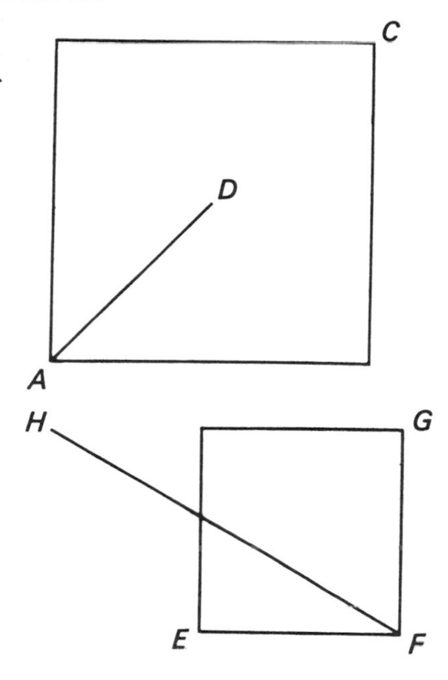

Now we will show how, having found one solution,[16] the other may be found without the help of the preceding rule. Let the constant of the equation be the product of AD and the square AC, and likewise let the sum of AD^2 and the square AC be the coefficient of x. According to Chapter VIII, AD will be the value of x. Let FH be such a line that, if one-half AD be added to it, the square of the sum will be equal to the square AC plus the square of one-half AD. I say that FH is the second solution. Since the cube plus the number, $AD \times AC$, is equal to the first power times the square AC plus AD^2, this makes the square EG which, when FH^2 is added to it, is equal to the square AC plus AD^2. Since, therefore, the square of the composite $(FH + \frac{1}{2}AD)$ is equal to the squares of CA and one-half AD, according to II, 4 of the *Elements*, by subtracting the common square of $\frac{1}{2}AD$ [from both sides of the equation], the square AC will be equal to $FH^2 + 2(FH \times \frac{1}{2}AD)$ and, therefore, to the rectangle $FH \times AD$ plus FH^2. Therefore, according to VI, 16 of the *Elements*, AB is a proportional between FH and the sum of FH and AD. But since the square EG plus the products of FH times itself and times AD produces as much as if it were added to the square AC, then certainly EG plus FH^2 is equal to AD^2 plus AC. By supposition, [the sum of] the square AC, AD^2 and the product $FH \times AD$ will be equal to [the sum of] the squares AC and EG. Hence, by subtracting from both sums AC^2, the square EG will be equal to the product $FH \times AD$ plus AD^2. According to VI, 16[17] [of the *Elements*], EF is a proportional between AD and the sum of AD and HF, wherefore, as has been shown,

[15] In 1545 and 1570 the lower right-hand corner of the first figure in the diagram is labeled E.

[16] 1570 and 1663 have "the greater solution."

[17] 1570 and 1663 omit this citation and have instead *per eandem*.

AB is likewise a proportional between FH and the sum of FH and AD, by the 34th proposition of the fifth book of our [work] on Euclid.[18] Since FH plus AD is the first quantity in both orders, the proportion

$$FH:AD = AB:EF$$

in duplicate and, therefore, according to VI, 17 of the *Elements*

$$FH:AD = AC:EG.$$

Hence, according to XI, 34 of the *Elements*, the body which is enclosed by FH and EG is equal to that enclosed by AD and AC and, therefore, to the constant of the equation. Whenever the square EG and HF^2 are equal to the x's, since [these are equal] to the square AC and AD^2, HF will, according to Chapter VIII of this book, be the value of x. Whence this rule:

Rule

Square one-half of the first[19] solution and multiply by three; subtract this from the coefficient of x and the square root of the remainder, minus one-half the first[20] solution, is the solution sought for.

Example:

$$x^3 + 60 = 46x$$

and one[21] solution for x is 6. To derive the other, square 3, one-half the first solution, which makes 9, and this multiplied by 3 is 27; subtract 27 from 46 and 19 remains; from the square root of this subtract 3, one-half the first solution, and you will have the second value of x as $\sqrt{19} - 3$. And by the same method, if you had found this, namely $\sqrt{19} - 3$, as the first solution the other would be 6.[22]

[18] 1570 and 1663 have "according to the 67th [proposition] of the *De Proportionibus* or the fifth [book] of this [work]." This proposition reads thus: *Si fuerint aliquot quantitates ab una quantitate, aliaeque totidem ab eadem analogae, erit proportio tertiae unius ordinis ad tertiam alterius, ut secundae ad secundam duplicata, et quartae ad quartam triplicata, quintae ad quintam duplicata, atque sic de aliis.*

[19] 1545 has *primae*; 1570 and 1663 have *maioris.*

[20] 1545 has *prioris*; 1570 and 1663 have *maioris.*

[21] 1545 has *altera*; 1570 and 1663 have *maior.*

[22] 1570 and 1663 omit this last sentence and add the following:

There are three corollaries to this rule. The first is that the solution for the cube equal to the first power and constant is equal to the [sum of] the two solutions for the cube and the same constant equal to the first power with the same coefficient. Thus, if

$$x^3 = 16x + 21$$

and the solution is $\sqrt{9\frac{1}{4}} + 1\frac{1}{2}$, there will be two solutions for

$$x^3 + 21 = 16x,$$

the sum of which is $\sqrt{9\frac{1}{4}} + 1\frac{1}{2}$. One is 3, the other $\sqrt{9\frac{1}{4}} - 1\frac{1}{2}$. From this corollary and the rules given in this chapter, there follows a second, namely, that from the multiplication and subtraction [*ex mutua multiplicatione et detractione*] of either solution for the cube and constant equal to the first power arises the other solution. Third, the solutions [*sic*] for the cube and constant equal to the first power compare with the solutions for the cube equal to the same first power and constant as *apotome* to *binomium*. These solutions for the case of the cube equal to the first power and constant are found thus: the roots of the individual remainders [i.e., the result of subtracting three times the squares of one-half the first solution from the coefficient of x] are the first half of the *apotome*, and one-half the first solution is the second part. In the example above, you have the accepted solutions in reverse order. Notice how the lowest is constituted and that it is evident that the larger is necessarily 3 and the smaller $\sqrt{9\frac{1}{4}} - 1\frac{1}{2}$. Explore, and you will discover.

$$\sqrt{9\frac{1}{4}} - 1\frac{1}{2}$$
$$3$$
$$\sqrt{9\frac{1}{4}} + 1\frac{1}{2}$$

CHAPTER XIIII

On the Cube Equal to the Square and Number

If the cube is equal to the square and constant, the equation can be changed into one of the cube equal to the first power and constant by the first method of conversion, which is from the whole to a part. (The second method is from the part to the whole, the third by the difference between parts, and the fourth by proportion.)

DEMONSTRATION

Let the cube AE, in the diagram for Chapter XII, equal $6AC^2 + 100$. And since AC^2 consists of AB^2 plus the gnomon surrounding it,

$$AC^3 = 6AB^2 + 6 \text{ gnomons} + 100.$$

The gnomon, however, consists of $BC^2 + 2(AB \times BC)$. Therefore

$$AC^3 = 6AB^2 + 6BC^2 + 6(AB \times 2BC) + 100.$$

But the product of AB and $2BC$ is $4x$, since AB equals x and BC equals 2. And $6BC^2$ equals $3BC^3$, since BC is one-third of 6. Therefore

$$AC^3 = 6AB^2 + 24x + 3BC^3 + 100.$$

It is evident that 24, the coefficient of x, consists of 6,[1] the coefficient of the square [AB^2], multiplied by 4, which is twice one-third the same number. And, on the other hand, it is also clear that

$$AC^3 = AB^3 + BC^3 + 3(AB \times BC^2) + 3(BC \times AB^2),$$

as has been shown in the first proposition of Chapter VI. Therefore

$$AC^3 = AB^3 + BC^3 + 6x^2 + 12x.$$

[1] 1570 and 1663 have 9.

Hence

$$AB^3 + BC^3 + 6x^2 + 12x = 6x^2 + 24x + 3BC^3 + 100.$$

It is also evident that the coefficient of x^2 is the same [on both sides of the equation] — that is, $3BC$[2] — and that BC is one-third the coefficient of x^2 and that the coefficient of x is the product of the coefficient of x^2 and one-third itself. For this [i.e., the coefficient of x] is always equal to three times the square of one-third [the coefficient of x^2]. Therefore, having subtracted from both sides BC^3 and $6x^2$ and $12x$ — that is, as many x's as result from multiplying the coefficient of x^2 by one-third itself — there remain

$$AB^3 = 100 + 12x + 2BC^3.$$

It is also evident that the number, 100, is the same as before; that the coefficient of x is the product of the coefficient of x^2 and one-third itself; and that $2BC^3$ is 16, since BC equals 2. Whence

$$AB^3 = 12x + 116.$$

Therefore, having found AB in accordance with the preceding rule, we add to it BC, which is one-third the coefficient of x^2, and AC results; and since, in seeking AB, we cubed one-third the coefficient of x, this one-third the coefficient of x is the square of one-third the coefficient of x^2. From this last simplification, this rule is derived:

Rule

Add the cube of one-third the coefficient of x^2 to one-half the constant of the equation and square the sum of these. From this square subtract the cube of the square of one-third the coefficient of x^2 and add or subtract the square root of the remainder to or from the sum[3] which was squared. You will then have a *binomium* and its *apotome*. Add their cube roots and add to them one-third the coefficient of x^2 and the whole which results is the value of x.[4]

[2] *quia est triplus ad BC.*

[3] The text has "one-half the sum."

[4] I.e., if $x^3 = bx^2 + N$,

$$x = \sqrt[3]{(b/3)^3 + N/2 + \sqrt{[(b/3)^3 + N/2]^2 - (b/3)^6}} + \sqrt[3]{(b/3)^3 + N/2 - \sqrt{[(b/3)^3 + N/2]^2 - (b/3)^6}} + b/3.$$

For example,

$$x^3 = 6x^2 + 20.$$

Add 8, the cube of 2, which is one-third of 6, to 10, one-half of 20, making 18, from the square of which, 324, subtract 64, the cube of the square of 2, which leaves 260, the square root of which add to or subtract from 18. You will then have $18 + \sqrt{260}$ and $18 - \sqrt{260}$, the sum of the cube roots of which, added to one-third the coefficient of x^2, is the value of x.

CHAPTER XV

On the Cube and Square Equal to the Number

DEMONSTRATION

In this case we convert by the second method. The difference is that the first method showed the addition of one-third the coefficient of x^2, while the second shows its subtraction. In the diagram for Chapter XII,[1] let

$$AB^3 + 6AB^2 = 100,$$

and assume that BC is equal to one-third the coefficient of x^2. Complete the cube of AC. Then

$$AC^3 = AB^3 + 6AB^2 + 12AB + BC^3,$$

from the first proposition of Chapter VI. Therefore in place of $AB^3 + 6AB^2$ put 100, for these are equal to 100. Then

$$AC^3 = 12AB + BC^3 + 100.$$

But $12AB$ differs from $12AC$ by $12BC$, and this 12, as has heretofore been shown, comes from three times BC^2; therefore $12BC$ equals $3BC^3$. Hence

$$AC^3 + 3BC^3 = 12AC + BC^3 + 100.$$

Now subtracting BC^3 alike from both sides,

$$AC^3\,[2] + 2BC^3 = 12AC + 100.$$

But $2BC^3$ is 16 and the coefficient of AC is equal to $3BC^2$, or one-third the coefficient of x^2 and, therefore, having found the value of AC, we subtract from it BC, one-third the coefficient of x^2, which leaves AB known. Hence the rule is as follows:

[1] 1570 and 1663 have Chapter XIII.

[2] The text has AB^3.

RULE

Cube one-third the coefficient of x^2, multiply the result by 2, and take the difference between this and the constant of the equation. [The result is the constant of a new equation.] Then multiply the square of one-third the coefficient of x^2 by 3 and you will have [the number of] y's which are equal to the cube and the constant, if twice the cube is greater than the constant of the equation, or [the number of] y's which with the constant, are equal to the cube, if twice the cube is less than the constant of the equation, or [the number of] y's which equal the cube, if the difference between these two numbers is zero. Having derived the solution [for this equation], subtract one-third the coefficient of x^2 from it and the remainder is the value of x.[3]

For example,

$$x^3 + 6x^2 = 100.$$

The cube of 2 is 8 which, being multiplied by 2, is 16. Subtract this from 100 and you will have

$$y^3 = 84 + 12y.$$

The y's are 12, or three times the square of 2, one-third of 6, the coefficient of x^2. The value of y, therefore, according to Chapter XII, is $\sqrt[3]{42 + \sqrt{1700}}$ +[4] $\sqrt[3]{42 - \sqrt{1700}}$. From this subtract 2, one-third of 6,[5] and the desired result, when

$$x^3 + 6x^2 = 100$$

will be this: $\sqrt[3]{42 + \sqrt{1700}} + \sqrt[3]{42 - \sqrt{1700}} - 2.$

Again, let

$$x^3 + 6x^2 = 25.$$

[3] I.e., if $x^3 + bx^2 = N$, and

$$(1)\ \text{if } 2\left(\frac{b}{3}\right)^3 > N, \text{ then } y^3 + \left[2\left(\frac{b}{3}\right)^3 - N\right] = 3\left(\frac{b}{3}\right)^2 y;$$

$$(2)\ \text{if } 2\left(\frac{b}{3}\right)^3 < N, \text{ then } y^3 = \left[N - 2\left(\frac{b}{3}\right)^3\right] + 3\left(\frac{b}{3}\right)^2 y; \text{ and}$$

$$(3)\ \text{if } 2\left(\frac{b}{3}\right)^3 = N, \text{ then } y^3 = 3\left(\frac{b}{3}\right)^2 y$$

and $x = y - \frac{b}{3}$.

[4] 1663 omits the indication of addition.

[5] 1570 has 9.

I subtract 16, twice the cube of one-third of 6, from 25, making 9. As before, then, this plus $12y$ is equal to y^3. The value of y, therefore, is $\sqrt{5\frac{1}{4}}$[6] $+ 1\frac{1}{2}$. Subtract 2 and the value sought for turns out to be $\sqrt{5\frac{1}{4}} - \frac{1}{2}$.

Again,

$$x^3 + 6x^2 = 16.$$

Subtract twice the cube of 2, namely 16, from 16, the constant, and nothing remains. Then take three times the square of this same one-third the coefficient of x^2. This is 12, the number of y's equal to y^3. Therefore y^2 equals 12, and y equals $\sqrt{12}$. Subtract 2, one-third of 6, and $\sqrt{12} - 2$ is left as the value of x.

Again,

$$x^3 + 6x^2 = 7.$$

Take the difference between 7 and 16, twice the cube of 2, and it is 9, and since twice the cube is greater than the constant of the equation and the coefficient of y is 12, as before, we will have

$$y^3 + 9 = 12y.$$

Therefore y equals 3 or $\sqrt{5\frac{1}{4}} - 1\frac{1}{2}$. Subtract 2, and the value of the [unknown in] $x^3 + 6x^2$ [$= 7$] will be either 1 or $\sqrt{5\frac{1}{4}} - 3\frac{1}{2}$, and this latter is negative, for $-3\frac{1}{2}$ is greater than $\sqrt{5\frac{1}{4}}$. Hence

$$6x^2 = 105 - \sqrt{9261}.$$

Truly x^3 is $\sqrt{9261} - 98$. If, therefore, x^3 and $6x^2$ are added together, they make exactly[7] 7, as is evident.

Corollary.[8] From this it is clear why the case of the cube and the constant equal to the square has not been demonstrated from the case of the cube and square equal to the constant. How, then, can the case of the cube and the number equal to the first power be demonstrated from the case of the cube equal to the first power and number? Since this case leads to the case of the cube and the number equal to the first power, it is better to proceed from the case of the cube and the number equal to the square directly to that of the cube and the number equal to the first power than to the same by way of the case of the cube and square equal to the number. This avoids a far longer and more confusing demonstration.

[6] The text has $\sqrt{5\frac{1}{3}}$.

[7] 1545 has *praecise*; 1570 and 1663 have *ad unguem*.

[8] 1570 and 1663 omit this word.

DEMONSTRATION

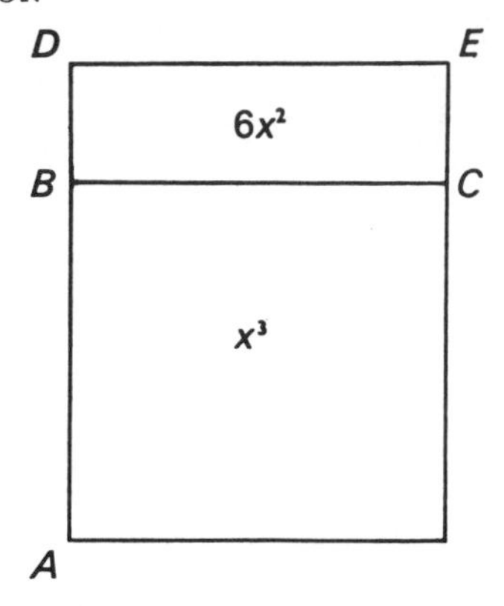

[There follows] another demonstration [of this rule] which is similar to our general [demonstration] in Chapter VII. It was discovered by Lodovico Ferrari.[9] Let the cube AC plus $6x^2$ (for example, CD) equal 100. Since, then, BD is the altitude of $6x^2$, BD will be 6. Therefore, assuming that AD is a certain square [y^2], AB will be $y^2 - 6$. Therefore the surface AC is $y^4 + 36 - 12y^2$, and this is the base of the body AE, wherefore the body

$$AE =^{10} y^6 + 36y^{2\,11} - 12y^4,$$

and this is equal to 100. Therefore 10, the square root of 100, equals $y^3 - 6y$,[12] [or] $\sqrt{y^6 + 36y^2 - 12y^4}$.[13] The value of y, therefore, is known and, being multiplied by itself, yields AD, since AD was postulated as y^2. Having subtracted BD, which is 6, from this, AB is left as the value of x.

RULE

The rule, therefore, is: Let the coefficient of x^2 be the coefficient of y which, with the square root of the given constant, is equal to the cube. Having arrived at a solution for this, square it and subtract from this the coefficient of x^2 or of y and the remainder is the value of x.[14]

For example,

$$x^3 + 6x^2 = 40.$$

You say, therefore, that

$$y^3 = 6y + \sqrt{40}.$$

Then the value of y, according to the appropriate rule, is $\sqrt[3]{\sqrt{10} + \sqrt{2}} + \sqrt[3]{\sqrt{10} - \sqrt{2}}$, the square of which is $\sqrt[3]{12 + \sqrt{80}} + \sqrt[3]{12 - \sqrt{80}}$[15] $+ 4$.

[9] 1545 has *Demonstratio alia Ludovici, similis nostrae universali, capituli 7ⁱ, et fuit inventa haec a Ludovico de Ferrariis*; 1570 and 1663 have *Demonstratio alia similis nostrae generali, capituli septimi inventa a Ludovico de Ferrariis.*

[10] 1663 omits any indication of equality.

[11] 1663 omits y^2.

[12] 1545 has *1 cub 6 co:.*

[13] 1663 repeats $-12y^4$.

[14] I.e., if $x^3 + bx^2 = N$, let $y^3 = by + \sqrt{N}$. Then $x = y^2 - b$.

[15] 1570 and 1663 have 50.

Subtract 6, the coefficient of y, from this, and the desired solution is found: $\sqrt[3]{12 + \sqrt{80}} + \sqrt[3]{12 - \sqrt{80}} - 2$. You would discover the same from the first rule of operation. The proof in the example of

$$x^3 + 3x^2 = 21$$

is this: The value of x derived from these rules is $\sqrt[3]{9\frac{1}{2} + \sqrt{89\frac{1}{4}}} + \sqrt[3]{9\frac{1}{2} - \sqrt{89\frac{1}{4}}} - 1$. The cube, therefore, is composed of seven parts:

$$12 - \sqrt[3]{4846\tfrac{1}{2} + \sqrt{23{,}487{,}833\tfrac{1}{4}}}\text{ [16] } - \sqrt[3]{4846\tfrac{1}{2} - \sqrt{23{,}487{,}833\tfrac{1}{4}}}$$

$$+ \sqrt[3]{46{,}041\tfrac{3}{4} + \sqrt{2{,}119{,}776{,}950\tfrac{7}{8}} - \sqrt{2{,}096{,}289{,}117\tfrac{9}{16}}\text{ [17] } - \sqrt{2{,}096{,}354{,}180\tfrac{13}{16}}\text{ [18]}}$$

$$+ \sqrt[3]{46{,}041\tfrac{3}{4} + \sqrt{2{,}096{,}354{,}180\tfrac{13}{16}} - \sqrt{2{,}096{,}289{,}117\tfrac{9}{16}} - \sqrt{2{,}119{,}776{,}950\tfrac{7}{8}}}$$

$$+ \sqrt[3]{256\tfrac{1}{2}\text{ [19] } + \sqrt{65{,}063\tfrac{1}{4}}} + \sqrt[3]{256\tfrac{1}{2} - \sqrt{65{,}063\tfrac{1}{4}}}.$$

The $3x^2$, moreover, are in seven parts after this fashion:

$$9 + \sqrt[3]{4846\tfrac{1}{2} + \sqrt{23{,}487{,}833\tfrac{1}{4}}} + \sqrt[3]{4846\tfrac{1}{2} - \sqrt{23{,}487{,}833\tfrac{1}{4}}}$$

$$- \sqrt[3]{256\tfrac{1}{2} + \sqrt{65{,}063\tfrac{1}{4}}} - \sqrt[3]{256\tfrac{1}{2} - \sqrt{65{,}063\tfrac{1}{4}}}$$

$$- \sqrt[3]{256\tfrac{1}{2} + \sqrt{65{,}063\tfrac{1}{4}}} - \sqrt[3]{256\tfrac{1}{2} - \sqrt{65{,}063\tfrac{1}{4}}}.$$

Then, adding the $3x^2$ to x^3, the six parts which are under the universal radical signs cancel each other and exactly 21 is left.[20]

[16] 1663 fails to indicate the universal character of the cube root.

[17] 1570 and 1663 have $\sqrt{20{,}096{,}286{,}117\frac{9}{16}}$.

[18] 1663 omits this last square root; it also adds, following the universal root, an extra expression: $\sqrt[3]{46{,}041\frac{3}{4} + \sqrt{2{,}096{,}354{,}180\frac{13}{16}}}$.

[19] 1570 and 1663 have $\sqrt{226\frac{1}{2}}$.

[20] 1570 and 1663 add at this point the following:

PROBLEM

There is a square column 36 cubits high and one cubit in breadth and depth. Its weight is exactly equal to that of another square column. If 6 cubits height are subtracted from the latter, the remainder will be a solid all the faces of which are square. Assuming, therefore, that the breadth of the column is x,

$$x^3 + 6x^2 = 36.$$

Therefore x equals $\sqrt[3]{16} + \sqrt[3]{4} - 2$, and this is the breadth of the base of the column. Its altitude, however, is 6 cubits more. Hence the altitude is $\sqrt[3]{16} + \sqrt[3]{4} + 4$.

CHAPTER XVI

On the Cube and Number Equal to the Square

RULE

This rule is clearly evident from the demonstration of the seventh chapter. The rule is: Multiply the cube root of the constant by the coefficient of x^2. This will produce a number of y's equal to the cube and the same constant. Having derived the solutions [for this equation], square the cube root of the constant and divide it by whatever values for y have been derived, and the results will be the values of x that are being sought.[1]

For example,

$$x^3 + 64 = 18x^2.$$

Multiply 18 by 4, the cube root of 64, making 72, the number of y's which are equal to $y^3 + 64$. The solutions for this are, according to its own rule, 8 and $\sqrt{24} - 4$. Divide 16, the square of 4, the cube root of 64, by these. The results are 2 and $\sqrt{96} + 8$, and these are the solutions.

DEMONSTRATION

Let one of the values of x be AB. I wish to have the other. I square AB — which is AC — and subtract AB from the coefficient of x^2, leaving AD. I now multiply AD by the sum of AB and one-fourth of AD; take the side of the resulting surface[2]; and add to it one-half of AD, making EF. This, I say, is the second value of x.

Construct EF^2 and assume that EG plus EF is equal to the sum of AB and AD. Since EF^2 is equal to the square of the side of the tetragon[3] plus the square of one-half AD plus the product[4] of the side of the

[1] I.e., if $x^3 + N = ax^2$, $a\sqrt[3]{N}y = y^3 + N$, and $(\sqrt[3]{N})^2/y = x$.

[2] *et superficiei productae sumatur latus quod in eam potest.*

[3] *producto ex tetragonali in se.*

[4] 1663 has *procto.*

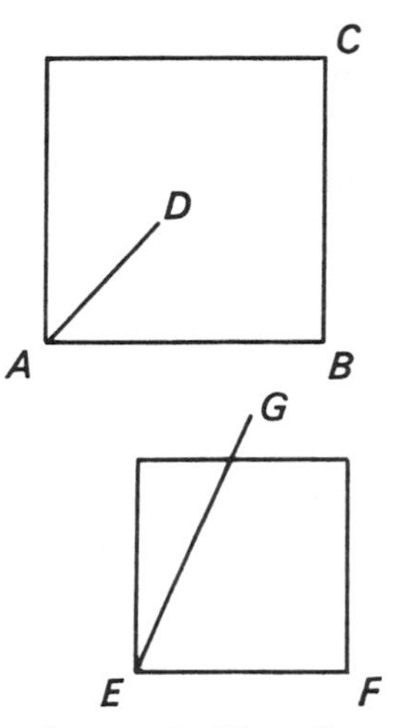

tetragon and AD, according to II, 4 of the *Elements*, EF^2 will be equal to AD times the sum of AB and one-half AD and the side of the tetragon. According to VI, 16 of the *Elements*, therefore, EF is a mean between AD and the sum of AB, the side of the tetragon, and one-half AD. Half of AD and the side of the tetragon, however, constitute EF, by supposition. EF, therefore, is a proportional[5] between AD and the sum of AB and EF. Again, the product of $(AB + AD)$ and $(AB + EF)$ is equal to the product of $(EF + EG)$ and $(AB + EF)$ since, by agreement, $EF + EG$ equals $AB + AD$, while AB and EF remain the same. The product of AD and $(AB + EF)$, however, is equal to EF^2, from the proofs. Therefore

$$AB(AB + EF) + EF^2 = (EF + EG)(EF + AB).$$

Now subtract from both sides alike EF^2 and

$$AB(AB + EF) = EG(AB + EF) + (EF \times AB).$$

Then, subtracting again the product common to both, $AB \times EF$, there remains

$$AD^2 = EG(AB + EF),$$

wherefore AB is a mean between EG and the sum of AB and EF. But, as was said, EF is truly a mean between AD and the sum of AB and EF. These are, therefore, three proportional[6] quantities in two orders, of which the first in both orders is the same, namely the sum of AB and EF. Therefore, according to the 34th proposition of the fifth book of our work on Euclid,[7]

$$EG:AD = AB^2:EF^2,^8$$

wherefore, according to VI, 17 of the *Elements*,

$$EG:AD = AC:EF^2;$$

and therefore, according to XI, 34 of the *Elements*, the body $AD \times AC$ is equal to the body $EG \times EF^2$. But AB was the value of x. Therefore,

[5] 1545 has *proportionalis*; 1570 and 1663 have *media . . . proportione*.

[6] 1545 has *proportionales*; 1570 and 1663 have *analogae*.

[7] So in 1545; 1570 and 1663 have "the 67th proposition of the fifth book of this work." See note 18, p. 108, *supra*.

[8] *EG ad AD, ut AB ad EF duplicata.*

the body $AD \times AC$ equals the constant of the equation, the sum of AD and AB having been assumed to be the coefficient of x^2 in accordance with the demonstration in Chapter VIII. Hence the product of EG and EF^2 equals the constant of the equation. Since, therefore, EF[9] $+ EG$ are equal to the coefficient of x^2 (this [sum being equal to] the sum of AB and AD) and since $GE \times EF^2$ is the constant of the equation, EF will, by Chapter VIII, be the value of x, as was to be proved.

Rule

The rule, therefore, is: Subtract the first solution from the coefficient of x^2 and multiply the remainder by the sum of the first solution and one-fourth this same remainder. Take the square root of the product. Add to it one-half the same remainder. The total is the value of x sought for.[10]

For example, let

$$x^3 + 24 = 8x^2$$

and let the known value of x be 2. Subtract 2 from 8, the coefficient of x^2, and 6 will be left. Multiply this by $3\frac{1}{2}$ (which is made up of 2, the first solution, and $1\frac{1}{2}$, which is one-fourth of 6, the remainder) and 21 results, to the square root of which add one-half the remainder,[11] which is 3,[12] making $\sqrt{21} + 3$[13] the solution sought for.

[9] 1570 and 1663 have CF.

[10] I.e., if $x^3 + N = ax^2$ and if $x = r, s$, then $s = \sqrt{(a - r)\left(r + \frac{a - r}{4}\right)} + \frac{a - r}{2}$.

[11] 1545 has *dimidium primae aestimationis*; 1570 and 1663 have *dimidium residui primae aestimationis*, apparently the result of inserting the correct word *residui* to correct 1545 and failing to strike out the incorrect words *primae aestimationis*.

[12] 1545 has 1, which is correct with the incorrect text just noted; 1570 and 1663 have 3.

[13] 1545 has $\sqrt{21} + 1$ which, again, is correct with the incorrect text but will not solve the problem given; 1570 and 1663 have $\sqrt{21} + 3$.

CHAPTER XVII

On the Cube, Square, and First Power Equal to the Number

Demonstration[1]

As an example, let

$$AB^3 + 6AB^2 + 20AB = 100.$$

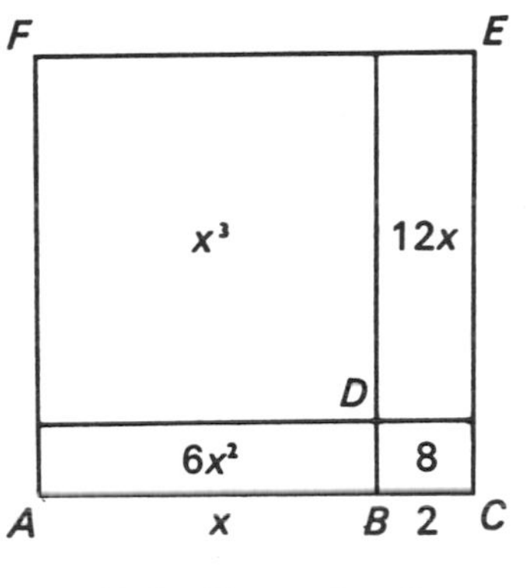

I add BC, which is 2, or one-third the coefficient of the second power, to AB and construct the whole cube of AC. This is made up of eight parts which, according to Chapter VI of this book, will be AB^3 (which is the surface FD and its altitude); BC^3, or 8, since BC equals 2; the [three] bodies on AD (which are equal to $6AB^2$); and the [three] bodies on DE (which are equal to $12AB$[2]). Since, therefore,

$$AB^3 + 6AB^2 + 20AB = 100,$$

add $8AB$, which is the difference between $20AB$ [and $12AB$], to AC^3, which equals $AB^3 + 6AB^2 + 12AB + BC^3$, and then

$$AC^3 + 8AB = 108,$$

for AC^3 is greater than the three bodies DA, DE, and DF[3] by the cube CD, which is 8. And since $8AB$ are less than the $8AC$ of the whole cube

[1] The lettering in the diagrams in 1570 and 1663 varies from that in 1545 which is used here. In 1570 an *O* replaces the *C* in the lower right-hand corner. In 1663 the upper left-hand corner is labeled *C*, the upper right-hand corner *F*, and the lower right-hand corner *O*.

[2] *et corpora de 12AB seu duodecuplo AB.*

[3] 1570 and 1663 omit *DF*.

by $8BC$[4] — BC being 2 — we add, accordingly, $8BC$ to both sides and make

$$AC^3 + 8AC = 124.$$

AC being known, therefore, from its own rule, we subtract BC, and this leaves AB.

Again let

$$AB^3 + 6AB^2 + 12AB = 100.$$

By adding BC^3 to both, AC^3 will equal 108 and AC, accordingly, is $\sqrt[3]{108}$, and AB will be 2 less than AC.

Let, again,

$$AB^3 + 6AB^2 + 2AB = 100.$$

Having added the remaining $10AB$ to complete the body DE and having added BC^3, AC^3 is equal to the additional $10AB$ plus 108. But $10AB$ are less than $10AC$ by $10BC$. We add, therefore, $10BC$ to both sides, making

$$AC^3 + 20 = 10AC + 108.$$

Subtract 20 from both sides and AC^3 remains equal to $10AC + 88$. Then, having found AC, subtract BC, and AB is necessarily left as known.

Rule

The common rule, therefore, is this: Cube one-third the coefficient of the square (which we show by this sign, *Tpq̃d*[5]) and add it to the constant. Then multiply the coefficient of the square by one-third itself, and the difference between this product and the coefficient of x is the number of y's to be added to the cube if the product is less than the given coefficient of x or to be added to the constant if the product is greater than the given coefficient of x.

If the difference between the product and the coefficient of x is zero, y^3 will be equal to the number first derived, and the cube root of this number, minus one-third the coefficient of x^2, is the value of x.

[4] *in 8BC seu octuplo BC.*

[5] T[ertia] p[ars] [numeri] quad[ratorum] = one-third the coefficient of x^2. This abbreviation is used very frequently in the text from here on.

If y^3 and the y's are equal to a number, [then in order to get the new constant], multiply the coefficient of y by one-third the coefficient of x^2 and add this product to the number first derived. You then have the cube and the first power with the coefficient already discovered equal to the constant just derived. From the solution for this subtract one-third the coefficient of x^2. The remainder is the value of x.

If the product is greater than the coefficient of x, multiply the difference — that is, the coefficient of y — by one-third the coefficient of x^2 and subtract the product from the whole number which you first derived. If nothing remains you have y^3 equal to the first power with the coefficient just derived and, by reducing to a lower power, you will have y^2 equal to the number and y will be the square root of the coefficient of y. From this subtract one-third the coefficient of x^2 and the remainder will be the value of x. But if, in subtracting the product of the coefficient of y and one-third the coefficient of x^2 from the aggregate [that you first derived], there is a remainder, this number plus y with the coefficient already given will equal y^3. Hence by subtracting one-third the coefficient of x^2 from the solution [for this equation], the remainder will be the value of x. But if the product of the coefficient of y and one-third the coefficient of x^2 is greater than the aggregate, the difference is the number which with y^3 is equal to the y's as already given, whence by subtracting one-third the coefficient of x^2 from the solution [for this equation] the remainder is the value of x.[6]

[6] I.e., if $x^3 + bx^2 + ax = N$, and

(1) if $\frac{b^2}{3} = a$, then $y^3 = \left(\frac{b}{3}\right)^3 + N$;

(2) if $\frac{b^2}{3} < a$, then $y^3 + \left(a - \frac{b^2}{3}\right)y = \left(a - \frac{b^2}{3}\right)\frac{b}{3} + \left(\frac{b}{3}\right)^3 + N$;

(3) if $\frac{b^2}{3} > a$, and

(i) if $\left(\frac{b}{3}\right)^3 + N = \left(\frac{b^2}{3} - a\right)\frac{b}{3}$, then $y^3 = \left(\frac{b^2}{3} - a\right)y$;

(ii) if $\left(\frac{b}{3}\right)^3 + N > \left(\frac{b^2}{3} - a\right)\frac{b}{3}$, then $y^3 = \left(\frac{b^2}{3} - a\right)y + \left(\frac{b}{3}\right)^3 + N - \left(\frac{b^2}{3} - a\right)\frac{b}{3}$;

(iii) if $\left(\frac{b}{3}\right)^3 + N < \left(\frac{b^2}{3} - a\right)\frac{b}{3}$, then $y^3 + \left(\frac{b^2}{3} - a\right)\frac{b}{3} - \left[\left(\frac{b}{3}\right)^3 + N\right] = \left(\frac{b^2}{3} - a\right)y$;

and $x = y - (b/3)$.

Corollary.[7] From this it is patent that this case resolves itself into five cases, which are here placed in the margin, and[8] that it cannot be resolved into more. In some of the following [chapters], the resolution turns into only three.[9] In all four-term cases, however, it is a common [characteristic] that one-third the coefficient of x^2 is always added to or subtracted from the value arrived at after they have been resolved into two or three terms, just as in this rule it is always subtracted, and it is also common in every case that the coefficient of y and the [second] constant are built up in the same way, just as in this instance the coefficient of y is the difference between the coefficient of x given in the four-term equation [10]and the product[10] of the coefficient of x^2 and one-third itself; and the constant of the equation into which it [i.e., the original equation] resolves itself is the difference between the product of the coefficient of y and one-third the coefficient of x^2 and the sum of the cube of one-third the coefficient of x^2 and the constant of the first equation.

$y^3 + cy = M$
$y^3 = M$
$y^3 = cy$
$y^3 = cy + M$
$y^3 + M = cy$

Problem I

Example: There is an entirely square body[11] which with its surfaces and edges equals 22. You say, therefore,

$$x^3 + 6x^2 + 12x = 22.$$

Therefore cube 2, one-third the coefficient of x^2, making 8; add this to 22, making 30; then multiply 6, the coefficient of x^2, by 2, one-third itself, making 12[12] and the difference between this and the coefficient of x is nothing, for the x's were also 12. We have, accordingly, y^3 equal to 30 and y equal to $\sqrt[3]{30}$. Subtract 2, one-third the coefficient of x^2, from this, and the value of x is $\sqrt[3]{30} - 2$.

Moreover, this is the proof: Add x^3, $6x^2$, and $12x$ and these make 22:

$$\begin{aligned} 6x^2 &= 24 + \sqrt[3]{194{,}400} - \sqrt[3]{414{,}720} \\ x^3 &= 22 - \sqrt[3]{194{,}400} + \sqrt[3]{51{,}840} \\ 12x &= \sqrt[3]{51{,}840} - 24. \\ \text{Total} &= 22. \end{aligned}$$

[7] 1570 and 1663 omit this word.
[8] 1545 and 1570 have *et*; 1663 has *si*.
[9] *In aliquibus autem sequentium resolutio fit in tria postrema tantum.*
[10] 1663 omits these words.
[11] *Est corpus quadratum undequaque.*
[12] 1570 and 1663 have 2.

PROBLEM II

Example of the second [form]: Find four numbers in continued proportion, the first of which is 3 and the remainder of which total 19. Let the second be x, the third will be $\frac{1}{3}x^2$, and the fourth will be $\frac{1}{9}x^3$.[13] Therefore

$$x + \tfrac{1}{3}x^2 + \tfrac{1}{9}x^3 = 19.$$

Turn these into whole numbers and you will have

$$x^3 + 3x^2 + 9x = 171,$$

for all are multiplied by 9. Add, therefore, to 171 the cube of one-third the coefficient of x^2, which is 1, and 172 results; then multiply 3, the coefficient of x^2, by one-third itself and this makes 3, the difference between which and 9, the coefficient of x, is 6, the coefficient of y. This plus y^3 equals a constant. Since the product [3] was less [than the coefficient of x, 9], multiply 6, the coefficient of y, by one-third the coefficient of x^2, making 6; add this to 172 and 178 results. Therefore

$$y^3 + 6y = 178$$

and y will be $\sqrt[3]{\sqrt{7929} + 89} - \sqrt[3]{\sqrt{7929} - 89}$.[14] From this subtract one-third the coefficient of x^2, which is 1, and you will have the second quantity or $\sqrt[3]{\sqrt{7929} + 89}$[15] $- \sqrt[3]{\sqrt{7929} - 89} - 1$, and from it you can derive the others.

Example of the third form:

$$x^3 + 6x^2 + x = 14.$$

Add the cube of 2, one-third the coefficient of x^2, which is 8, to 14, and 22 results. Then multiply 6, the coefficient of x^2, by 2, one-third itself, making 12,[16] the difference between which and the coefficient of x is 11. [This is] the number of y's equal to y^3 and a constant. Since the number produced, 12, is greater than the coefficient of x, multiply 11 by 2, one-third the coefficient of x^2, making 22, the difference between which and the first sum [8 + 14] is nothing. Therefore we will have y^3 equal to $11y$, and therefore y^2 equals 11, wherefore y equals $\sqrt{11}$. From this subtract 2, one-third the coefficient of x^2, and this makes x

[13] The text has $\frac{1}{4}x^3$.

[14] The text has a plus between the two universal radicals rather than a minus.

[15] 1663 omits +89.

[16] 1663 has 21.

equal to $\sqrt{11} - 2$. You took the difference between the numbers and not the sum, because y equals the cube rather than the cube plus y equaling the number, as in the preceding example.

Problem III

Example of the fourth form: An oracle ordered a prince to build a sacred building whose space should be 400 cubits, the length being six cubits more than the width, and the width three cubits more than the height. These quantities are to be found.

Let the altitude be x; the width will be $x + 3$; and the length will be $x + 9$. Multiplied in turn, you will have

$$x^3 + 12x^2 + 27x = 400.$$

Add the cube of 4, one-third the coefficient of x^2, which is 64, to 400, and 464 results. Multiply 12, the coefficient of x^2, by one-third itself, making 48, the difference between which and 27 is 21, the number of y's which equal the cube and constant. Therefore multiply 21[17] by 4, one-third the coefficient of x^2, and 84 results; take the difference between this and 464, which is 380; add this to y, since the first sum of numbers was greater than the second number produced; and you will have

$$y^3 = 21y + 380.$$

Therefore y will be $\sqrt[3]{190 + \sqrt{35,757}}$[18] $+ \sqrt[3]{190 - \sqrt{35,757}}$. From this subtract 4, one-third the coefficient of x^2, and you will have the altitude and, having this, by adding 3 and 9 to it, you will derive the width and length also, as you see:

Altitude: $\sqrt[3]{190 + \sqrt{35,757}} + \sqrt[3]{190 - \sqrt{35,757}} - 4$

Width: $\sqrt[3]{190 + \sqrt{35,757}} + \sqrt[3]{190 - \sqrt{35,757}} - 1$

Length: $\sqrt[3]{190 + \sqrt{35,757}} + \sqrt[3]{190 - \sqrt{35,757}} + 5$

Example of the fifth form:

$$x^3 + 6x^2 + 2x = 3.$$

Add 8, the cube of one-third the coefficient of x^2, to 3, making 11. Then multiply 6[19] by one-third itself, and 12 results, the difference between

[17] 1663 has 12.

[18] 1570 and 1663 have $\sqrt{35,750}$.

[19] 1570 and 1663 have 9.

which and 2, the coefficient of x, is 10, the coefficient of y. Multiply by 2, one-third the coefficient of x^2, making 20, the difference between which and 11 is 9, the number which with y^3 equals $10y$, since the second product is greater than the totaled number. (I call the second product that which derives from the coefficient of y, now known, multiplied by one-third the coefficient of x^2.) Therefore the value of y, when

$$y^3 + 9 = 10y,$$

is either 1 or $\sqrt{9\frac{1}{4}} - \frac{1}{2}$. Subtract, therefore, 2, one-third the coefficient of x^2, and the two desired solutions are obtained. One is $\sqrt{9\frac{1}{4}} - 2\frac{1}{2}$, the other -1.

CHAPTER XVIII

On the Cube and First Power Equal to the Square and Number

DEMONSTRATION

Using the same figure let, for example,

$$AC^3 + 33AC = 6AC^2 + 100,$$

and let AC^3 be divided into its parts, assuming BC to be one-third the coefficient of x^2 or 2. Then,

$$AC^3 = AB^3 + BC^3 + 6AB^2 + 12AB,$$

and

$$33AC = 33AB + 33BC,$$

[the latter of] which equal 66, since BC equals 2. Therefore,

$$AC^3 + 33AC = AB^3 + BC^3 + 6AB^2 + 45AB + 66.$$

These, accordingly, are equal to $6AC^2 + 100$. But $6AC^2$, since AC has been divided at B, are equal to $6AB^2 + 6BC^2 + (12 \times AD$, the surface), according to II, 4 of the *Elements*. But AD equals $2AB$,[1] since BC equals 2. Therefore $12AD$ will equal $24AB$, wherefore

$$6AB^2 + 6BC^2 + 24AB + 100 = AB^3 + BC^3 + 6AB^2 + 45AB + 66.$$

However, BC^3 equals 8, and $6BC^2$ equals 24. Therefore,

$$6AB^2 + 24AB + 124 = AB^3 + 6AB^2 + 45AB + 74.$$

Accordingly, by making a like subtraction from both sides — namely $6AB^2 + 24AB + 74$ — there remain

$$AB^3 + 21AB = 50.$$

It is, accordingly, clear that by deriving AB from this equation and adding BC, which is 2, to it, AC will be found. It is also clear that if the [number of] x's which were on the same side as the cube [in the original

[1] *2 positiones.*

equation] had been equal to the product [of the coefficient of x^2 and one-third the same], we would have had a cube equal to a number alone [in the reduced equation]; and if the [number of] x's which were on the same side as the cube had been less [than that product], we would have had a first power on one side and a cube on the other. In this case, if the constant had been equal to a [certain] other [number], the first power would equal the cube, but if it [i.e., the new first power] were smaller [than the new cube] we would have had the first power and number equal to the cube and, if greater, the first power equal to the cube and a number, according to the same demonstration as in the preceding chapter.[2]

RULE

The rule, therefore, is that you always set up the coefficient of y first, as in the preceding chapter. [In order to do so] you multiply the coefficient of x^2 by one-third itself, and the difference between this product and the coefficient of x is the number of y's. If this is nothing, we will have a cube equal to a number. If, on the other hand, the product is less than the coefficient of x, the difference will be the number of y's which, with a cube, equal a number. And if the product is greater, we will have the y's equal to a cube.[3]

[2] *Manifestum est autem, quod ubi positiones, quae cum cubo erant, essent aequales productis, haberemus cubum aequalem numero tantum, et ubi positiones quae cum cubo erant, essent pauciores, haberemus res ex una parte, et cubum ex alia, et tunc si numerus qui est cum cubo, foret aequalis alteri, essent positiones aequales cubo, et si essent minor, haberemus res et numerum aequales cubo, et si maior, haberemus res aequales cubo et numero, ex eadem demonstratione, velut in praecedente capitulo.*

This opaque passage is troublesome in a number of respects: (1) One is the incompleteness and, therefore, the ambiguity of *productis* and *alteri* which, to make sense, need to be amplified. (2) Another is the use of the plural *productis*, which has been translated as though it were a singular. (3) A third is the meaning of *positiones* the third time it occurs in the passage. The first two times it occurs it quite clearly refers to the x terms in the original equation, just as *res* refers to the y terms in the reduced equation. Here, however, *positiones* appears to be used in place of *res*, to the confusion of the reader. (4) Finally, there is the *qui est cum cubo* clause, which has been omitted in the translation since, if the cube referred to is supposed to be the one in the original equation, the discussion has no place in this chapter and, if it is supposed to be the one in the reduced equation, the statement is immediately contradicted by the following clauses. Fortunately Cardano restates the whole passage in somewhat clearer form in setting out his rule and this, plus the examples, enables us to read it as here given.

[3] I.e., if $x^3 + ax = bx^2 + N$, then (1) if $\frac{b^2}{3} - a = 0$, $y^3 = M$; (2) if $\frac{b^2}{3} < a$, $y^3 + \left(a - \frac{b^2}{3}\right)y = M$; (3) if $\frac{b^2}{3} > a$, $y^3 = \left(\frac{b^2}{3} - a\right)y$, with or without M on one side or the other.

In this last case, if the numbers [i.e., the constant of the original equation plus its increment and the product of the coefficient of x and one-third the coefficient of x^2] are the same, y^3 will be equal to the y's [alone]. But if the product of the coefficient of x and one-third the coefficient of x^2 is less than the constant of the equation with its increment, y^3 will be equal to the y's and a number. And if the product of the coefficient of x and one-third the coefficient of x^2 is greater than the constant and its increment, the y's will be equal to y^3 and a number.[4]

This number, moreover, is derived thus: Multiply the coefficient of x by one-third the coefficient of x^2 and take the difference between[5] this product and the sum of the constant of the equation and twice the cube of one-third the coefficient of x^2.[6] The difference will be the number which is to be added either to y^3 or to y, as is proper in the circumstances,[7] or the number which is equal to y^3 if there are no y's. Then, having derived the solution [for y], add or subtract one-third the coefficient of x^2 as you will learn from the examples, and you will have the solution for x.

Example 1:

$$x^3 + 12x = 6x^2 + 25.$$

Multiply 6 by 2, one-third of itself, making 12, the difference between which and the coefficient of x is nothing. Therefore y^3 will equal a number. Multiply, accordingly, 12, the coefficient of x, by 2, one-third the coefficient of x^2, making 24. Subtract this from 41, the sum of 16, twice the cube of 2, and 25, the constant of the equation, leaving 17, which is equal to y^3. Y, therefore, is $\sqrt[3]{17}$.[8] Add 2, one-third the coefficient of x^2, to this, and the value of x appears as $\sqrt[3]{17} + 2$.

[4] I.e., if $N + n = \frac{ab}{3}$, $y^3 = dy$; if $N + n > \frac{ab}{3}$, $y^3 = dy + M$; if $N + n < \frac{ab}{3}$, $y^3 + M = dy$, d being equal to $\frac{b^2}{3} \sim a$ as in the preceding footnote.

[5] 1545 has *cum*; 1570 and 1663 have *ab*.

[6] I.e., $M = \frac{ab}{3} \sim \left[N + 2\left(\frac{b}{3}\right)^3\right]$, $2\left(\frac{b}{3}\right)^3$ being the increment (n) spoken of in the text but otherwise unidentified.

[7] So in 1545; 1570 and 1663 spell it out more fully: "the number which is to be added to the cube if the ["first," in 1570 only] product is greater than the sum or to the first power if it is less."

[8] 1545 has $\sqrt[3]{16}$.

Example 2: A bankrupt merchant bargains for the repayment of three-fourths of his debts in three years in such proportion that, if the agreement were for repayment of $\frac{19}{27}$ of what he owes, he would return $\frac{9}{27}$ the first year, $\frac{6}{27}$ the second, and $\frac{4}{27}$ the third, and the remainders would be in continued proportion with the principal sum.[9] He now wishes to know the portion to be paid each year if he repays three-fourths. We propose, in order to avoid fractions, that his capital is 4. He wishes, then, to repay 3. Assume that he returns x the first year; therefore the second year he returns $x - \frac{1}{4}x^2$, and the third year $x - \frac{1}{2}x^2 + \frac{1}{16}x^3$. Therefore in the three years he returns $3x + \frac{1}{16}x^3 - \frac{3}{4}x^2$, and this is supposed to be 3. Hence, raising these terms to integers by multiplying the cube by 16, you will have

$$x^3 + 48x = 12x^2 + 48.$$

Multiply 12 by 4, one-third of itself, making 48; then the difference [between this and the coefficient] of x is zero, and a cube will equal a number. Now multiply 48, the coefficient of x, by 4, one-third the coefficient of x^2, and 192 results. From this subtract 176, the sum of twice the cube of 4, and 48, the constant, and 16 remains. This is equal to y^3. Therefore the value of y is $\sqrt[3]{16}$, which subtract from 4, one-third the coefficient of x^2, and the solution sought is $4 - \sqrt[3]{16}$. He returns, accordingly, in the first year $4 - \sqrt[3]{16}$, in the second $\sqrt[3]{16} - \sqrt[3]{4}$, and in the third $\sqrt[3]{4} - 1$. The remainders of these are in continued proportion with 4 and, added together, they [the three complex terms] equal 3. This is the converse of the first example.[10] The remainders themselves are $\sqrt[3]{16}$, $\sqrt[3]{4}$, and 1.

Example 3:

$$x^3 + 15x = 6x^2 + 24.$$

Multiply 6 by one-third itself, and 12 results, the difference between which and 15, the coefficient of x, is 3. Since the product is the smaller, y^3 and $3y$ will equal a number. Multiply, therefore, 15, the coefficient of x, by 2, one-third the coefficient of x^2, making 30; subtract from 40,

[9] *ut residua sint in eadem proportione, cum residuo capitali.* I read *et* for *ut* and omit *residuo* as inconsistent with the problem as it is worked out and with the conclusions stated at the end.

[10] Meaning that here we have $2\left(\frac{b}{3}\right)^3 + N$ less than $\frac{ab}{3}$, whereas there we had $2\left(\frac{b}{3}\right)^3 + N$ greater than $\frac{ab}{3}$.

the sum of 24 and twice the cube of one-third the coefficient of x^2, and 10 remains. Therefore,

$$10 = y^3 + 3y,$$

and the value of y is $\sqrt[3]{\sqrt{26} + 5} - \sqrt[3]{\sqrt{26} - 5}$.[11] Add to this 2, one-third the coefficient of x^2, and you will have the desired solution.

Example 4:

$$x^3 + 15x = 6x^2 + 10.$$

Again I have the cube and $3y$ equal to a number, and the product-number will be 30, as before. The sum of twice the cube of 2, one-third the coefficient of x^2, and 10, the constant of the equation, is 26, and the difference is therefore 4. Since, accordingly,

$$y^3 + 3y = 4,$$

the value of y is 1. And since the product-number is greater than the sum — it is 30, which is greater than 26 — we subtract 1, the solution for the second equation, from 2, one-third the coefficient of x^2, and 1 is left as the desired solution for

$$x^3 + 15x = 6x^2 + 10.$$

Hence it is evident in this case, where the cube and first power are equal to the number, that if the difference between the numbers were nothing — as if we put 14 in place of 10 — the solution would be one-third the coefficient of x^2, or 2, since the derived equation would give us nothing to add or subtract, for

$$y^3 + 3y = 0.$$

Example 5:

$$x^3 + 10x = 6x^2 + 4.$$

Multiply the coefficient of x^2 by one-third itself, as before, and 12 is the result; the difference between this and the coefficient of x is 2, and since the product is greater than the coefficient of x, $2y$ will equal y^3. To derive the constant for this equation, multiply 10, the coefficient of x, by 2, one-third the coefficient of x^2, making 20, the difference between which and 20, the sum of twice the cube of one-third the coefficient of x^2 and 4, is nothing. Therefore, we will have no number, but y^3 will equal, as we said, $2y$. Therefore, by depression, y^2 will equal

[11] The text omits the internal radical sign.

2, and therefore y is $\sqrt{2}$. Add this to or subtract it from one-third the coefficient of x^2; you will have the true solution of the equation, either $2 + \sqrt{2}$, or $2 - \sqrt{2}$, and it may even be 2, and thus this case has three solutions.

Example 6: Let

$$x^3 + 21x = 9x^2 + 5.$$

Now, as before, I multiply 9 by 3, one-third of itself, making 27, the difference between which and 21, is 6, which will be the number of y's that are equal to y^3, since 27, the product, is greater than 21, the coefficient of x. Add, therefore, 54, twice the cube of one-third the coefficient of x^2, to 5, the constant of the equation, thus obtaining 59, the difference between which and 63, the product of the coefficient of x and one-third the coefficient of x^2, is 4. Therefore, since the product is greater than the sum, we add the number to the cube and have

$$y^3 + 4 = 6y,$$

the $6y$ having already been derived. Therefore, there are three solutions for this equation: the first is 2, the second $\sqrt{3} - 1$, and the third, which is false, is $-(\sqrt{3} + 1)$.[12] Add these to 3, one-third the coefficient of x^2, and you will have these true solutions: 1st, 5; 2d, $2 + \sqrt{3}$; 3d, $2 - \sqrt{3}$.

Example 7:

$$x^3 + 26x = 12x^2 + 12.$$

Multiply 12, the coefficient of x^2, by one-third itself, which is 4, and 48 results, the difference between which and 26, the coefficient of x, is 22. Since the product is greater than the coefficient of x, the y's will be equal to y^3. Then multiply 26, the coefficient of x, by 4, one-third the coefficient of x^2, making 104; subtract this from 140, the sum of twice the cube of one-third the coefficient of x^2, and 12, the constant, and 36 is the result, which number is to be added to the y's, because the sum is greater than the product. Contrary to the preceding example, therefore,

$$y^3 = 22y + 36.$$

Therefore, there will be three solutions, the first of which is $\sqrt{19} + 1$ and is true; the second is false, $-(\sqrt{19} - 1)$; and the third is also

[12] *tertia ficta m: ℞ 3 p: 1.*

false and is -2. Add these individually to one-third the coefficient of x^2, and you will have three solutions which I show thus: 1st, $5 + \sqrt{19}$; 2d, $5 - \sqrt{19}$; 3d, 2.

From this it is evident that the coefficient of x^2, in the three examples in which there are three solutions for x, is always the sum of the three solutions: as in the fifth case, where $(2 + \sqrt{2}) + 2$[13] $+ (2 - \sqrt{2})$ make up 6, the coefficient of x^2; and as in the sixth example, $5 + (2 + \sqrt{3}) + (2 - \sqrt{3})$ make up 9, the coefficient of x^2; and as in the seventh example, $(5 + \sqrt{19}) + (5 - \sqrt{19}) + 2$ are 12, the coefficient of x^2. Hence knowing two such solutions, the third always emerges. The reason for this appears at the beginning of this book. It is also evident that whenever we arrive [at a case in which] the y's are separated from the cube, whether the number is joined to the y's or to y^3, three solutions always appear. The reason for this is likewise given at the same place above where we spoke of true and false solutions. And it is also evident that all these methods can always be carried back to addition, for a minus that is added acts the same way as a plus that is subtracted, for it is clear that to subtract 4 from 12 is the same as to add -4 to 12. Either way, the result is 8.

Here is another demonstration of this rule, discovered by Lodovico Ferrari, showing more clearly the reason for these operations:[14]

Another Demonstration

Let

$$x^3 + 100x = 6x^2 + 10$$

and assume that AB is the value of x and BC one-third the coefficient of x^2. Let, moreover, AG be equal to BC. Then GB is the difference between BC and AB. GB^3, moreover, is the difference between $BC^3 + 3(BC \times AB^2)$ and $AB^3 + 3(AB \times BC^2)$, in accordance with the sixth chapter. Now, by agreement,

$$AB^3 + 100AB = 6AB^2 + 10.$$

But

$$6AB^2 = 3(BC \times AB^2).$$

[13] 1663 omits this term.

[14] This statement in 1545 is reserved for the end of the next demonstration in 1570 and 1663. In place of it appears the following: "From this it appears that the coefficient of x^2 is divided three ways and having one solution the sum of the others is known."

Therefore the sum of $3(BC \times AB^2)$ and BC^3, which is 8, is less than $AB^3 + 100AB$ by 2. I say -2, because BC^3, which is added to $6AB^2$, would have to be 10 but is only 8. And $AB^3 + 100AB$ is greater than $AB^3 + 3(AB \times BC^2)$, which is $12AB$, by $88AB$. The difference, therefore, between $BC^3 + 3(BC \times AB^2)$ and $AB^3 + 3(AB \times BC^2)$ is $88AB - 2$. Therefore this difference, as we said, is equal to GB^3.

x

A B G C

Assume, now, that BG is x. Then GC or AB will be $2 - x$, which quantity taken 88 times, as was said, is equal to $BG^3 + 2$. Therefore

[15] $$BG^3 + 2 + 88BG = 176,$$ [15]

whence

$$BG^3 + 88BG = 174,$$

wherefore, if you subtract this solution, BG, from BC, which is one-third the coefficient of x^2, or 2, you will have the quantity, AB, which was being sought.[16]

Another Demonstration

Assume again that

$$x^3 + 5x = 6x^2 + 10$$

and let EF be x and DE one-third the coefficient of x^2. Then the difference between DE and EF will be EH. From a demonstration similar to the foregoing, then,

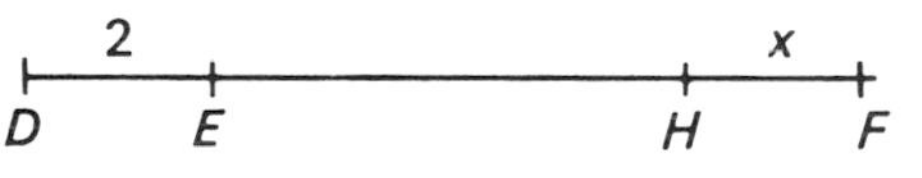

$$EH^3 = 7EH + 16.$$

Having found the solution [for this equation], add to it HF, one-third the coefficient of x^2, which is 2, and the sought-for solution, EF, will be found. Nor in this case, [need] I add a word, since the demonstration is similar to the foregoing and the operation in this part is clearer in our demonstration.

[15] 1663 omits this passage.

[16] 1570 and 1663 add at this point: *Et haec demonstratio fuit inventa a Ludovico Ferrario et ostendit clarius rationem supradictarum operationum.*

RULE

The rule derived from this demonstration is that if the coefficient of x is equal to the product of the coefficient of x^2 and one-third itself, cube one-third the coefficient of x^2, add the cube root of the difference between this and the constant of the equation to one-third the coefficient of x^2 in case the cube is less than the number, or subtract it if it is greater, and the whole is the value of x.[17] It is evident, moreover, that when the cube of one-third the coefficient of x^2 and the number are equal we neither add nor subtract, but one-third the coefficient of x^2 will itself be the solution for x.

For example,

$$x^3 + 12x = 6x^2 + 8.$$

Now since multiplying 6, the coefficient of x^2, by one-third itself makes 12, the coefficient of x exactly, cube 2, one-third the coefficient of x^2, making 8, the difference between which and the constant is zero; therefore x equals 2, or one-third the coefficient of x^2.

And if

$$x^3 + 12x = 6x^2 + 9,$$

since the cube [of one-third the coefficient of x^2] is less than[18] the constant we subtract 8, the cube of one-third the coefficient of x^2, from 9, leaving 1, the cube root of which is 1. This we add to one-third the coefficient of x^2, since the cube of one-third the coefficient of x^2 is less than the constant of the equation, and this gives 3 as the value of x.

And, for the same reason, if

$$x^3 + 12x = 6x^2 + 7,$$

subtract 7 from 8, the cube of one-third the coefficient of x^2, leaving 1, the cube root of which is 1. Subtract this from 2, one-third the coefficient of x^2, leaving 1 as the value of x.

But if the coefficient of x is greater than the product of the coefficient of x^2 and one-third itself, the difference will be the coefficient of y, as in the first demonstration and, according to its rules, it is multiplied by one-third the coefficient of x^2, the cube of one-third the coefficient of x^2 is added to this, and the difference between this sum and the

[17] I.e., if $x^3 + ax = bx^2 + N$, and if $a = b^2/3$, $x = b/3 + \sqrt[3]{N - (b/3)^3}$ if $(b/3)^3 < N$, or $b/3 - \sqrt[3]{(b/3)^3 - N}$ if $(b/3)^3 > N$.

[18] The text has "equal to."

constant of the [original] equation is the constant of the [new] equation of the cube and the y's. If the difference is zero, the value of x is one-third the coefficient of x^2. And if the constant of the equation is less than the sum, subtract the solution [for the new equation] from, or if greater add it to, one-third the coefficient of x^2, and what results will be the value of x.[19]

For example,

$$x^3 + 20x = 6x^2 + 24.$$

Multiply 6 by 2, one-third of itself, and 12 is produced, the difference between which and 20, the coefficient of x, is 8, the number of y's which, with the cube, are equal to a number. Multiply, therefore, 8,[20] the coefficient of y, by 2, one-third the coefficient of x^2, making 16. Add to this 8, the cube of one-third the coefficient of x^2, making 24. The difference between this and 24, the constant, is nothing. Therefore the value of x is one-third the coefficient of x^2, namely 2.

Again, let

$$x^3 + 20x = 6x^2 + 15.$$

We will have, therefore, $y^3 + 8y$ as before. To derive the number, multiply as before 8, the coefficient of y, by 2, one-third the coefficient of x^2, making 16. Add to this the cube of one-third the coefficient of x^2, making 24. Subtract 15, and 9 remains. Therefore,

$$y^3 + 8y = 9,$$

and the value of y is 1. Subtract this from 2, one-third the coefficient of x^2, leaving the true solution as 1. You subtracted, moreover, because 15, the constant of the equation, is less than the sum of the cube and the product, which is 24. If you notice closely, [you will see that] this is

[19] I.e., if $x^3 + ax = bx^2 + N$, and if $a > \frac{b^2}{3}$, then

$$y^3 + \left(a - \frac{b^2}{3}\right)y = N \sim \left[\left(a - \frac{b^2}{3}\right)\frac{b}{3} + \left(\frac{b}{3}\right)^3\right].$$

Then, if

$$(1)\ N = \left[\left(a - \frac{b^2}{3}\right)\frac{b}{3} + \left(\frac{b}{3}\right)^3\right],\quad x = \frac{b}{3};$$

$$(2)\ N < \left[\left(a - \frac{b^2}{3}\right)\frac{b}{3} + \left(\frac{b}{3}\right)^3\right],\quad x = \frac{b}{3} - y;$$

$$(3)\ N > \left[\left(a - \frac{b^2}{3}\right)\frac{b}{3} + \left(\frac{b}{3}\right)^3\right],\quad x = \frac{b}{3} + y.$$

[20] 1570 and 1663 omit this figure.

done by the method [given] in the first part of the rule, where the coefficient of x is the product of the coefficient of x^2 and one-third itself.

Again,

$$x^3 + 20x = 6x^2 + 33.$$

As before, you will again have $y^3 + 8y$ equal to the difference between 24, the sum, and 33, the constant of the equation, whence

$$y^3 + 8y = 9,$$

and the value of y will be 1. Adding to this one-third the coefficient of x^2, since the constant, 33, is greater than the sum, 24, the solution for x will be 3.

But if the coefficient of x is less than the product of the coefficient of x^2 and one-third itself, the difference nevertheless will be the coefficient of y, as before, but the y's will not be added to the cube but, on the contrary, will be equated to it. Then multiply this coefficient of y by one-third the coefficient of x^2 and add the product to the constant of the equation, and the difference between this sum and the cube of one-third the coefficient of x^2 is the constant of the second equation. If, then, there is no difference, the cube will equal the first power, and the square root of the coefficient of y plus one-third the coefficient of x^2 is the value of x. But if the sum is greater than the cube, the difference will be the number which, with the y's, is equal to y^3. Hence, having derived the solution for this, add to it one-third the coefficient of x^2, and this will give the true solution. But if the cube is greater than the sum, the difference will be the number which with the cube equals the y's. Hence, having obtained the solution for this, add to it one-third the coefficient of x^2 and the result is the true solution for x and is multiplex, as we have shown in our rule.[21] Moreover, whatever pertains to the rule is ours.

[21] I.e., if $x^3 + ax = bx^2 + N$, and if $a < \frac{b^2}{3}$, and if

(1) $\left(\frac{b^2}{3} - a\right)\frac{b}{3} + N = \left(\frac{b}{3}\right)^3$, then $y^3 = \left(\frac{b^2}{3} - a\right)y$;

(2) $\left(\frac{b^2}{3} - a\right)\frac{b}{3} + N > \left(\frac{b}{3}\right)^3$, then $y^3 = \left(\frac{b^2}{3} - a\right)y + \left[\left(\frac{b^2}{3} - a\right)\frac{b}{3} + N\right] - \left(\frac{b}{3}\right)^3$;

(3) $\left(\frac{b^2}{3} - a\right)\frac{b}{3} + N < \left(\frac{b}{3}\right)^3$, then $y^3 + \left(\frac{b}{3}\right)^3 - \left[\left(\frac{b^2}{3} - a\right)\frac{b}{3} + N\right] = \left(\frac{b^2}{3} - a\right)y$,

and $x = y + \frac{b}{3}$.

For example,

$$x^3 + 9x = 6x^2 + 2.$$

In this case, the coefficient of y will be 3. Multiply this by 2, one-third the coefficient of x^2, making 6. Add this to 2, the constant of the equation, and 8 results. The cube of one-third the coefficient of x^2, however, is 8 and the difference is nil. Therefore y^3 equals $3y$, and the value of y is $\sqrt{3}$, and x, accordingly, is $2 + \sqrt{3}$.

Again,

$$x^3 + 9x = 6x^2 + 4.$$

We shall, as before, have y^3 equal to $3y$. For the number, multiply 3, the coefficient of y, by 2, one-third the coefficient of x^2, making 6. Add 4, the constant of the equation, making 10. Subtract 8, the cube of one-third the coefficient of x^2, and 2 is the result. This is to be added to the y's, because the sum is greater than the cube of one-third the coefficient of x^2. Hence

$$y^3 = 3y + 2,$$

and y will be 2. By adding 2, one-third the coefficient of x^2, 4 results as the true solution [for x].

Still again,

$$x^3 + 21x = 9x^2 + 5.$$

Therefore there will be 6 y's in the second equation. Since 9, the coefficient of x^2, multiplied by 3, one-third of itself, produces 27, multiply 6, the coefficient of y, by 3, one-third the coefficient of x^2, making 18. Add 5 to this, making 23. The difference between this and the cube of one-third the coefficient of x^2, is 4, and since the sum is less than the cube, therefore

$$y^3 + 4 = 6y.$$

The solution, accordingly, is 2 or $\sqrt{3} - 1$, or, as the false one, $-(\sqrt{3} + 1)$. If, therefore, you add to these 3, one-third the coefficient of x^2, you will have as the solutions sought for 5, and $2 + \sqrt{3}$, and $2 - \sqrt{3}$,[22] by all of which it is true that

$$x^3 + 21x = 9x^2 + 5.$$

[22] The text gives 5, $4 + \sqrt{3}$, and $2 + \sqrt{3}$ as the solutions.

CHAPTER XIX

On the Cube and Square Equal to the First Power and Number

DEMONSTRATION

Let, for example,

$$AB^3 + 6AB^2 = 20AB + 200,$$

and assume that BC is 2, one-third the coefficient of AB^2. AC will then be $AB + 2$, and its cube will be $AB^3 + 6AB^2 + 12AB + 8$. It was proposed, however, that

$$AB^3 + 6AB^2 = 20AB + 200.$$

Let, therefore, $20AB + 200$ take the place of $AB^3 + 6AB^2$, and this makes

$$AC^3 = 32AB + 208.$$

And since $32AB$ are less than $32AC$ by $32BC$, let $32BC$ be added to both sides. Hence

$$32AC + 208 = AC^3 + 64,$$

for so much [i.e., 64] is $32BC$. Subtract 64 from both sides, and

$$AC^3 = 32AC + 144.$$

Then, having arrived at the solution for this, subtract BC, one-third the coefficient of AB^2, from it and AB will be left.

RULE

The rule, therefore, is: Multiply the coefficient of x^2 by one-third itself, add the product to the coefficient of x, and the sum will be the coefficient of y. Then multiply this number by one-third the coefficient of x^2 and take the difference between this product and the sum of the constant of the equation and the cube of one-third the coefficient of x^2.

If this is nothing, the cube will be equal to the first power. If, however, the product is less than the sum, the difference is the constant which, with the first power, is equal to the cube. But if the product is greater than the sum, the difference is the number which, with the cube, equals the first power. Then, having derived the solution [for this], subtract[1] one-third the coefficient of x^2 from it and the remainder is the true solution for x that was to be found.[2]

For example,

$$x^3 + 6x^2 = 20x + 56.$$

Multiply 6 by 2, one-third of itself, making 12. Add this to 20, and 32 results. Multiply 32 by 2, one-third the coefficient of x^2, making 64. Add 8, the cube of one-third the coefficient of x^2, to 56, the constant of the equation, and 64 results. The difference between the product and the sum is zero. The first power, therefore, equals the cube; wherefore, by depression, y^2 equals 32, and y equals $\sqrt{32}$. Hence the true solution is $\sqrt{32} - 2$.

Again,

$$x^3 + 6x^2 = 20x + 112.$$

Multiply 6 by 2, as before, making 12 which, added to 20, makes 32, the coefficient of y. Multiply this by 2, one-third the coefficient of x^2, and 64 results. Subtract from 120, the sum of the cube of one-third the coefficient of x^2 and the constant, and 56[3] remains as the number which, with $32y$, is equal to y^3. Therefore y equals $\sqrt{29} + 1$. Subtract one-third the coefficient of x^2 and the value of x remains as $\sqrt{29} - 1$.

Again,

$$x^3 + 6x^2 = 20x + 21.^4$$

[1] 1570 and 1663 omit this word.

[2] I.e., if $x^3 + bx^2 = ax + N$, and

(1) if $N + \left(\frac{b}{3}\right)^3 = \left(\frac{b^2}{3} + a\right)\frac{b}{3}$, then $y^3 = \left(\frac{b^2}{3} + a\right)y$;

(2) if $N + \left(\frac{b}{3}\right)^3 > \left(\frac{b^2}{3} + a\right)\frac{b}{3}$, then $y^3 = \left[N + \left(\frac{b}{3}\right)^3\right] - \left[\left(\frac{b^2}{3} + a\right)\frac{b}{3}\right] + \left(\frac{b^2}{3} + a\right)y$;

(3) if $N + \left(\frac{b}{3}\right)^3 < \left(\frac{b^2}{3} + a\right)\frac{b}{3}$, then $y^3 + \left[\left(\frac{b^2}{3} + a\right)\frac{b}{3}\right] - \left[N + \left(\frac{b}{3}\right)^3\right] = \left(\frac{b^2}{3} + a\right)y$,

and $x = y - \frac{b}{3}$.

[3] 1663 has 65.

[4] The text has 41.

You will have, therefore, as before y with a coefficient of 32 and 35[5] as the constant in the second equation, for by subtracting 29,[6] the sum of the [original] constant and 8, the cube of one-third the coefficient of x^2, from 64, the product of 32 and one-third the coefficient of x^2, 35[5] remains. Since, then, the product is greater than the sum,

$$35^{5} + y^3 = 32y,$$

and y will be 5, or $\sqrt{13\frac{1}{4}} - 2\frac{1}{2}$, or, as a false solution, $-(\sqrt{13\frac{1}{4}} + 2\frac{1}{2})$.[7] Subtract 2, one-third the coefficient of x^2, from these and you will have 3 as the true solution and, for the two false ones, $-4\frac{1}{2} + \sqrt{13\frac{1}{4}}$ and $-4\frac{1}{2} - \sqrt{13\frac{1}{4}}$, as we said in the first chapter.[8]

[5] The text has 15.

[6] The text has 49.

[7] *vel ficta* ℞ *13¼ p: 2½.*

[8] The statement of the problem has been altered, as indicated in the foregoing footnotes, to fit the solutions given.

CHAPTER XX

On the Cube Equal to the Square, First Power, and Number

Demonstration

Again, for example, let

$$AC^3 = 6AC^2 + 5AC + 88,$$

and assume that BC is one-third the coefficient of AC^2, or 2. It is evident, therefore, that

$$AC^3 = 6AB^2 + 12AB + AB^3 + BC^3.$$

The latter, therefore, are equal to $6AC^2 + 5AC + 88$. Strike out, then, the common BC^3 — that is, 8 — and there remain

$$AB^3 + 6AB^2 + 12AB = 6AC^2 + 5AC + 80.$$

But $6AC^2$ are greater than $6AB^2$ by the six gnomons of AB^2, which are $24AC - 6BC^2$ and the latter [i.e., $6BC^2$] are 24. Therefore,

$$6AB^2 + 29AC^{[1]} + 56 = AB^3 + 6AB^2 + 12AB.$$

Subtract $6AB^2$, therefore, as common to both sides, and there remain

$$29AC + 56 = AB^3 + 12AB.$$

But $29AC$ are greater than $29AB$ by $29BC$ and, therefore, by 58, since BC equals 2. Therefore add this number to the constant and

$$29AB + 114^{[2]} = AB^3 + 12AB.$$

Again subtract $12AB$ from both sides, and

$$17AB + 114 = AB^3.$$

Having derived the solution for this equation, add BC to it.

[1] 1570 and 1663 have $92AC$.

[2] 1570 has 144.

Rule

Hence the rule is: Multiply the coefficient of x^2 by one-third itself. Add the product to the coefficient of x, and the sum will be the number of y's equal to the cube. To find the number, moreover, multiply the coefficient of y by one-third the coefficient of x^2 and add the product to the constant of the equation. From this subtract the cube of one-third the coefficient of x^2 and the remainder is the number which, with the y's, equals y^3. Having derived the solution for this, add to it one-third the coefficient of x^2 and you will have the true answer.[3]

Problem

An example is this problem: A certain man lent 1728[4] *aurei* on compound interest,[5] as they say, or *sub usura rediviva*, on the condition that he should receive in the third year as capital and interest as much as one-half the capital plus one-half of what [the debtor] owed him at the end of the first year plus one-half of what he owed him at the end of the second year if he then retained the money and wished to release it from this same interest. Assume that the first year's due is $144x$[6]; the second will be $12x^2$ and the third year's x^3. This last will equal the sum of the halves of the other years. Therefore

$$x^3 = 6x^2 + 72x + 729.$$[7]

[3] I.e., if $x^3 = bx^2 + ax + N$, then $y^3 = \left(\frac{b^2}{3} + a\right)y + \left(\frac{b^2}{3} + a\right)\frac{b}{3} + N - \left(\frac{b}{3}\right)^3$, and $x = y + \frac{b}{3}$.

[4] 1570 and 1663 have 2728.

[5] *ad caput anni*. On this phrase, see David Eugene Smith, *History of Mathematics* (Boston, 1925), II, 565, and Florence Edler, *Glossary of Mediaeval Terms of Business* (Cambridge, Mass., 1934), p. 62, where it is discussed under the phrase *capo d'anno*. See also Cardano's *Ars Magna Arithmeticae*, where, in the 35th problem, the same phrase occurs as a contrast to *simplex: et quod reditus capitalis esset simplex et non ad caput anni*, and Chapter LVII of his *Practica Arithmeticae* (*Opera Omnia*, IV, 101), where he explains it more fully thus: *Redditibus autem ad caput anni est ut reditus primi anni adiiciatur capitali, et ex toto accipias reditum secundi anni, sub eadem portione, et ita adiicias reditum, secundi anni capitali, et totum fiet capitale pro tertio anno et ita de singulis exemplum, dedi tibi 100. livras ad caput anni, ad 10. pro 100. igitur dices qui dat 10. pro 100. facit redditum esse* $\frac{1}{10}$ *capitalis pro anno, igitur primo ad 100. adde* $\frac{1}{10}$, *fiunt librae 110. quibus pro secundo anno adde* $\frac{1}{10}$ *capitalis, fiunt 121. quibus adde pro tertio anno* $\frac{1}{10}$ *totius aggregati et fit 133* $\frac{1}{10}$, *et ita pro quarto anno adde* $\frac{1}{10}$ *partem fient 146* $\frac{41}{100}$, *et ita vides quod redditus iungitur capitali, et quod plus crescit census quam in simplici redditu.*

[6] *Pone igitur quod in capite primi anni haberet 144 res.*

[7] So in the text, though the statement of the problem would suggest 864 instead.

Accordingly, multiply 6, the coefficient of the square, by 2,[8] one-third itself, making 12. Add 72 to this and 84 results, the coefficient of y. Multiply 84 by 2, one-third the coefficient of x^2, making 168. Add this to 729, making 897. Subtract 8, the cube of one-third the coefficient of x^2, leaving 889. Hence

$$y^3 = 84y + 889.$$

The solution for this, therefore, will be $\sqrt[3]{444\frac{1}{2} + \sqrt{175{,}628\frac{1}{4}}} + \sqrt[3]{444\frac{1}{2} - \sqrt{175{,}628\frac{1}{4}}}$.[9] To this add 2, one-third the coefficient of x^2, and you have as the desired solution $\sqrt[3]{444\frac{1}{2} + \sqrt{175{,}628\frac{1}{4}}} + \sqrt[3]{444\frac{1}{2} - \sqrt{175{,}628\frac{1}{4}}}$[9] + 2. The cube of this is the sum of money which was owing to him the third year. Hence subtracting 1728, you will have his profit by proportional numbers.[10]

[8] 1570 has 1.

[9] 1570 and 1663 have $\sqrt{175{,}928\frac{1}{4}}$ in both these places.

[10] 1545 has *per terminos proportionales*; 1570 and 1663 have *per terminos analogos*.

CHAPTER XXI

On the Cube and Number Equal to the Square and First Power

Demonstration

Let

$$x^3 + 100 = 6x^2 + 24x,$$

and let the cube be AC^3 and let BC be one-third the coefficient of x^2. Now,

$$AC^3 = AB^3 + 6AB^2 + 12AB + BC^3.$$

BC^3 is 8. So we will have

$$AB^3 + 6AB^2 + 12AB + 108 = 6AC^2 + 24AC.$$

But $6AB^2$ are less than $6AC^2$ by six times the gnomon ADE, and $24AB$ are less than $24AC$ by $24BC$. Therefore

$$AB^3 + 6AB^2 + 12AB + 108 = 6AB^2 + 6 \text{ gnomons } ADE + 24AB + 48,$$

for $24BC$ are 48. Therefore by subtracting from both sides $6AB^2 + 12AB + 48$, there will be

$$AB^3 + 60 = 6 \text{ gnomons } ADE + 12AB.$$

The 6 gnomons ADE, however, are $24AB + 24$ — this because either of the surfaces AD or DE is $2x$, for BD is 2, and BC^2 is 4. Hence

$$36AB + 24 = AB^3 + 60.$$

Subtract 24 from both sides and

$$AB^3 + 36 = 36AB.$$

Hence, knowing AB, we add to it BC, which is one-third the coefficient of x^2, and the solution will flow from this.

Rule

The rule, therefore, is: Multiply the coefficient of x^2 by one-third itself, add the product to the coefficient of x, and the coefficient of y will result. Multiply this by one-third the coefficient of x^2 and take the difference between this product and the sum of the constant of the equation and the cube of one-third the coefficient of x^2. If this is nothing, the y's will equal y^3. If, however, the product is greater than the sum, the difference will be the number which, with the first power, equals the cube. And if the sum is greater than the product, the difference is the number which, with the cube, is equal to the first power. Hence, having derived the solution [for this reduced equation], add it to one-third the coefficient of x^2 and from this will arise the true solution.[1]

Remember, moreover, that when the original equation turns into an equation of the cube equal to the first power and a number, the true solution of the latter will be added to one-third the coefficient of x^2 and the smaller of the fictitious solutions will [likewise] be added negatively. Thus you will have both [true] solutions for the equation of the cube and the number equal to the first power and square, although that of the cube equal to the first power and constant has only one true solution.

Example:

$$x^3 + 64 = 6x^2 + 24x.$$

Multiply 6, the coefficient of x^2[2] by one-third itself, or 2, making 12. Add this to 24 and 36 results as the coefficient of y. This multiplied by 2, one-third the coefficient of x^2, is 72. The cube of 2 is 8 and this, added to 64, the constant of the equation, is also 72. Therefore, since the difference between these numbers is nothing, we will have y^3 equal to $36y$. Hence y^2 equals 36, and y will equal 6, in accordance

[1] I.e., if $x^3 + N = bx^2 + ax$, and

(1) if $N + \left(\frac{b}{3}\right)^3 = \left(a + \frac{b^2}{3}\right)\frac{b}{3}$, then $y^3 = \left(a + \frac{b^2}{3}\right)y$;

(2) if $N + \left(\frac{b}{3}\right)^3 < \left(a + \frac{b^2}{3}\right)\frac{b}{3}$, then $y^3 = \left(a + \frac{b^2}{3}\right)y + \left(a + \frac{b^2}{3}\right)\frac{b}{3} - \left[N + \left(\frac{b}{3}\right)^3\right]$;

(3) if $N + \left(\frac{b}{3}\right)^3 > \left(a + \frac{b^2}{3}\right)\frac{b}{3}$, then $y^3 + N + \left(\frac{b}{3}\right)^3 - \left(a + \frac{b^2}{3}\right)\frac{b}{3} = \left(a + \frac{b^2}{3}\right)y$;

and $x = y + \frac{b}{3}$.

[2] The text has "coefficient of x.

with the simple rule. Add to this 2, one-third the coefficient of x^2, making 8, the solution for x.

Again,

$$x^3 + 128 = 6x^2 + 24x.$$

Six multiplied by 2, as before, makes 12. To this add 24, making 36, the coefficient of y. Multiply 36 by one-third the coefficient of x^2 and 72 results. The difference between this and 136, the sum of 128, the constant of the equation, and 8, the cube of one-third the coefficient of x^2, is 64, which number is to be added to the cube, since 136, the sum, is greater than 72, the product. Therefore

$$y^3 + 64 = 36y,$$

and y equals 2 or $\sqrt{33} - 1$. Add these to one-third the coefficient of x^2 and the true solutions become 4 and $\sqrt{33} + 1$.

Again, let

$$x^3 + 9 = 6x^2 + 24x.$$

Multiply, as before, 6 by 2, one-third of itself, making 12. Add this to 24,[3] the coefficient of x, making 36, the coefficient of y, as before. Then multiply 36 by 2, one-third the coefficient of x^2, and 72 results, the difference between which and 17, the sum of 8, the cube of one-third the coefficient of x^2, and 9, the constant of the equation, is 55. Therefore, since the product is greater than the sum, we add 55 to the first power and we will have

$$y^3 = 36y + 55.$$

For this the true solution is $\sqrt{17\frac{1}{4}} + 2\frac{1}{2}$, the greater false one is -5, and the smaller false one is $-(\sqrt{17\frac{1}{4}} - 2\frac{1}{2})$[4] or, that you may see it more clearly, $2\frac{1}{2} - \sqrt{17\frac{1}{4}}$. Add, therefore, this solution and likewise the true one to one-third the coefficient of x^2, which is 2, and you will have the solutions sought for: one is $4\frac{1}{2} + \sqrt{17\frac{1}{4}}$, the other $4\frac{1}{2} - \sqrt{17\frac{1}{4}}$.[5]

[3] 1570 and 1663 have 34.

[4] The text has *m: v:* ℞ *27¼ m: 2½*.

[5] *Quaere* why Cardano does not also give -3 as a solution.

CHAPTER XXII

On the Cube, First Power, and Number Equal to the Square

Demonstration

Again, let

$$AC^3 + 4AC + 16 = 6AC^2$$

and let BC be one-third the coefficient of x^2 as before. We resolve, then, AC^3, which is equal to $AB^3 + 6AB^2 + 12AB + BC^3$ — BC^3 being equal to 8 — and all this plus $4AC + 16$ will be equal to $6AC^2$. Therefore, since $4AC$ is equal to $4AB + 4BC$ or $+8$,

$$AB^3 + 6AB^2 + 16AB + 32 = 6AC^2.$$

However, as was demonstrated,

$$6AC^2 = 6AB^2 + 24AB + 24.$$

Therefore

$$AB^3 + 6AB^2 + 16AB + 32 = 6AB^2 + 24AB + 24.$$

Subtract $6AB^2 + 16AB + 24$ from both sides and there remain

$$AB^3 + 8 = 8AB.$$

Hence, knowing AB, add to it BC, one-third the coefficient of x^2, and this makes known AC, which is the value of x.

Again,

$$AC^3 + 4AC + 1 = 6AC^2.$$

Therefore, as before,

$$6AC^2 = 6AB^2\,[1] + 24AB + 24.$$

[1] 1570 and 1663 have $6AD^2$.

But

$$AC^3 + 4AC + 1 = AB^3 + 6AB^2 + 16AB + 17,$$

wherefore, by subtracting the common terms, $6AB^2 + 16AB + 17$, the remainders will be equal. That is

$$AB^3 = 8AB + 7.$$

Hence as before, knowing AB you have AC by adding BC, one-third the coefficient of x^2.

Rule

The rule, therefore, is: Multiply the coefficient of x^2 by one-third itself and subtract the coefficient of x from the product. If this cannot be done the case is impossible of a true solution. Whatever remainder there is will be the coefficient of y. Then multiply the coefficient of x by one-third the coefficient of x^2 and add the product to the constant of the equation. Take the difference between this sum and twice the cube of one-third the coefficient of x^2. If this difference is nothing, y^3 is equal to the y's alone. But if twice the cube of one-third the coefficient of x^2 is greater, the difference is the number which is to be added to the y's; and if twice the cube is less than the sum, the difference is the number which is to be added to y^3. Thence, having discovered the solution, add to it one-third the coefficient of x^2 in order to derive the true solution.[2]

For example,

$$x^3 + 4x + 8 = 6x^2.$$

Multiply 6 by 2, one-third itself, making 12. Subtract 4, making 8, the coefficient of y. Multiply also 4, the coefficient of x, by 2, one-third the coefficient of x^2, which makes 8. Add this to 8, the constant of the equation, and 16[3] is produced. There is no difference between this

[2] I.e., if $x^3 + ax + N = bx^2$, and

(1) if $2\left(\frac{b}{3}\right)^3 = \left(\frac{ab}{3} + N\right)$, then $y^3 = \left(\frac{b^2}{3} - a\right)y$;

(2) if $2\left(\frac{b}{3}\right)^3 > \left(\frac{ab}{3} + N\right)$, then $y^3 = \left(\frac{b^2}{3} - a\right)y + 2\left(\frac{b}{3}\right)^3 - \left(\frac{ab}{3} + N\right)$;

(3) if $2\left(\frac{b}{3}\right)^3 < \left(\frac{ab}{3} + N\right)$, then $y^3 + \left(\frac{ab}{3} + N\right) - 2\left(\frac{b}{3}\right)^3 = \left(\frac{b^2}{3} - a\right)y$;

and $x = y + \frac{b}{3}$.

[3] 1570 and 1663 have 26.

and twice the cube of one-third the coefficient of x^2, which is also 16. Therefore y^3 equals $8y$, and y equals $\sqrt{8}$. To this add 2, one-third the coefficient of x^2, giving $\sqrt{8} + 2$ as the true solution.

Again,

$$x^3 + 4x + 16 = 6x^2.$$

Multiply 6 by 2, one-third the coefficient of x^2, as before, and 12 is produced. Subtract 4, the coefficient of x, from this, leaving 8 as the coefficient of y. Multiply 4, the coefficient of x, by 2, one-third the coefficient of x^2, making 8. Add this to 16, the constant of the equation, and 24 results. By subtracting 16, which is twice the cube of one-third the coefficient of x^2, from this, 8 will remain. Therefore we add 8 to y^3, since the sum is greater than twice the cube of one-third the coefficient of x^2, and this makes

$$y^3 + 8[4] = 8y,$$

wherefore y equals 2, or $\sqrt{5} - 1$. Therefore, by adding 2, one-third the coefficient of x^2, the true solution appears as 4, or $\sqrt{5} + 1$.

Again,

$$x^3 + 4x + 1 = 6x^2.$$

As before, y will have a coefficient of 8 and, having multiplied the coefficient of x, which is 4, by 2, one-third the coefficient of x^2, 8 results. To this add 1, the constant of the equation, making 9. Twice the cube of one-third the coefficient of x^2 is 16. Hence the difference is 7 and, since twice the cube is greater than the sum,

$$8y + 7 = y^3.$$

Therefore y equals $\sqrt{7\frac{1}{4}} + \frac{1}{2}$ or, as a false solution, the smaller value of y will be -1. To these add 2,[5] one-third the coefficient of x^2. You will then have two true solutions, namely 1, and $\sqrt{7\frac{1}{4}} + 2\frac{1}{2}$.[6]

You will remember, moreover, what we said in the preceding chapter, especially this: that when the [principal] equation is reduced to one of the cube equal to so many y's, [then][7] the false solution does not differ from the true one in number. Therefore for a second solution

[4] This figure is omitted in 1570 and 1663.

[5] 1570 and 1663 have 3.

[6] The text has $\sqrt{7\frac{1}{4}} + \frac{1}{2}$.

[7] The text has *quia*.

[for the principal equation], inasmuch as nothing is added to or subtracted from one-third the coefficient of x^2, this same one-third the coefficient of x^2 will be a true solution in both cases.[8] Thus the solution for

$$x^3 + 4x + 8 = 6x^2$$

will be either $\sqrt{8} + 2$, or 2, and in the preceding chapter the solution for

$$x^3 + 64 = 6x^2 + 24x$$

will be either 8, as was said, or 2 — that is, one-third the coefficient of x^2 — and this [is so] because all additions and subtractions have to be made to or from one-third the coefficient of x^2.

[8] I.e., in the case of $x^3 + ax + N = bx^2$ and in the case of $x^3 + N = bx^2 + ax$.

CHAPTER XXIII

On the Cube, Square, and Number Equal to the First Power

DEMONSTRATION

Let

$$x^3 + 6x^2 + 4 = 41x$$

and let it [the cube] be AB^3. To AB I add BC, one-third the coefficient of x^2. Hence

$$AC^3 = AB^3\,[1] + 6AB^2 + 12AB + 8.$$

Put $41AB$ in place of $AB^3 + 6AB^2 + 4$. Then

$$AC^3 = 53AB + 4$$

— 4 being the difference between BC^3 and 4, the constant of the first equation. To complete $53AC$, therefore, add $53BC$, and

$$AC^3 + 106 = 53AC + 4.$$

Subtract 4 from both sides, and

$$AC^3 + 102 = 53AC.$$

Then, having found the solution for AC, subtract BC, one-third the coefficient of x^2, from it and AB will thus be known, and this is x itself.

RULE

The rule, therefore, is: Multiply the coefficient of x^2 by one-third itself; add the product to the coefficient of x, and this gives the coefficient of y. From this subtract the square of one-third the coefficient of x^2, multiply the remainder by one-third the coefficient of x^2, and

[1] 1570 and 1663 have AE^3.

add the whole product to the constant of the equation, whence will arise the number which, with the cube, is equal to the first power already determined. Thence, from the solutions for this equation, subtract one-third the coefficient of x^2 and the remainders are the solutions sought for.[2] [For illustration of this rule] one example suffices:

$$x^3 + 6x^2 + 12 = 31x.$$

Multiply 6, the coefficient of x^2, by 2, one-third of the same, making 12. Add this to 31, making 43, the coefficient of y. From this subtract 4, the square of one-third the coefficient of x^2, leaving 39 which, multiplied by 2, one-third the coefficient of x^2, makes 78. Add this to 12, the constant of the equation, and 90 results. Hence

$$y^3 + 90 = 43y.$$

Therefore y equals 5 or $\sqrt{24\frac{1}{4}} - 2\frac{1}{2}$. Subtract 2, one-third the coefficient of x^2, and you will have as the true solutions 3 and $\sqrt{24\frac{1}{4}} - 4\frac{1}{2}$. By either of these, it is true that

$$x^3 + 6x^2 + 12 = 31x.$$

You must remember, however, that all the solutions for these cases are obtained always by subtracting[3] from the true and false solutions of the equations into which they are resolved one-third the coefficient of x^2. Then, provided that a number remains, even [if] that which is subtracted[3] is a pure negative, this remainder is a true solution for x. They can also be resolved into other cases of four powers when this is convenient.

[2] I.e., if $x^3 + bx^2 + N = ax$, $y^3 + \left[\frac{b^2}{3} + a - \left(\frac{b}{3}\right)^2\right]\frac{b}{3} + N = \left(\frac{b^2}{3} + a\right)y$. and $x = y - \frac{b}{3}$.

[3] The text has "adding" and "added" in these two places.

CHAPTER XXIIII

On the Forty-Four Derivative Equations

DEMONSTRATION

Now, let, for example,

$$x^6 + 6x^4 = 100$$

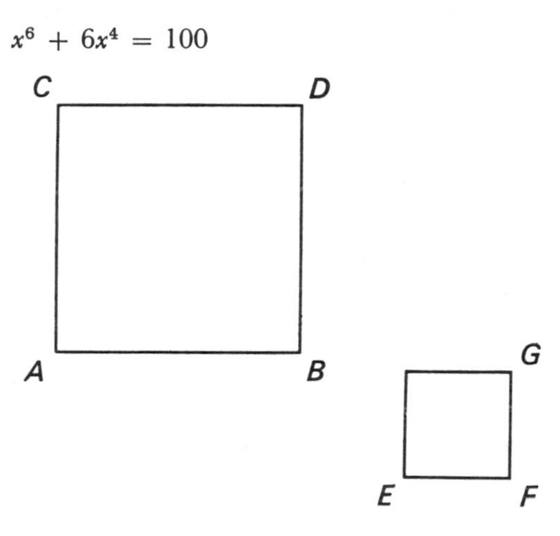

and let x^6 be the body *ABCD*, having a height of *AB*, which will, therefore, be a square, since it is the side of a body[1] *ABCD*, which is supposed to be the cube of a square. It is evident, then, that the surface *ABCD* is the square of a square, since *AB* is now supposed to be a square. Therefore, by supposition,

$$6ABCD \text{ (the surface)} + ABCD \text{ (the body)} = 100.$$

Assume, therefore, that *AB* is x. Then the body *ABCD* will be x^3 and the surface *ABCD* will be x^2. It has been supposed also, however, that *ABCD*, the body, plus 6*ABCD*, the surface, is equal to 100. Therefore,

$$AB^3 + 6AB^2 = 100.$$

Wherefore *AB* will be known according to the proper rules. But *AB*, in the first problem, was x^2. Therefore the value of x^2, in the first problem, when

$$x^6 + 6x^4 = 100$$

will be known, since it is the same as the value of x in the second. And we wish the value of x in the first case; x, however, is always the root

[1] *latus cubi cũ corporis.*

of a square. Therefore $\sqrt{AB}$, the solution found for the second problem, is the value of x in the first, as was proposed.

By the same reasoning, if we are given

$$x^6 + 6x^3 = 100,$$

the body $ABCD$ will [represent] x^6 and AB will be x^2. If we assume another square surface equal to this — I suggest EFG — it will be a case of

$$6(EF \times EFG) + ABCD \text{ (the body)} = 100.$$

But let the body EFG be x. Since, therefore, EF is $\sqrt{AB}$, by agreement, EF^3 will be $(\sqrt{AB})^3$. Therefore the body $ABCD$ is the square of the body $EF \times EG$. Assuming, therefore, the body $ABCD$ to be x^2, EF^3 will be x and six times this will be $6x$. Now

$$6EF^3 + ABCD \text{ (the body)} = 100.$$

The bodies do not change but remain the same, and $6EF^3$ is $6x$, and the body $ABCD$ is x^2. Therefore,

$$x^2 + 6x = 100.$$

Therefore x is known — that is, EF^3 — but since EF is the cube root[2] of its own cube, EF will be known. This is the cube root of the solution [just] discovered. And since EF was x in the first place, and since it is $\sqrt{AB}$, and AB is supposed to be x^2, postulating the body $ABCD$ as the cube of a square, therefore, [I say,] — postulating the body $ABCD$ as x^6 — EF will be x. And note that it is the cube root of the solution derived in the second case that we are looking for.

From this the rules for all the derivative equations are evident, for we have demonstrated generally the 16 primitive composite cases, and these are they:

1. $x^2 = ax + N$	9. $N = x^3 + bx^2$
2. $ax = x^2 + N$	10. $x^3 = bx^2 + ax + N$
3. $N = x^2 + ax$	11. $bx^2 = x^3 + ax + N$
4. $x^3 = ax + N$	12. $N = x^3 + bx^2 + ax$
5. $ax = x^3 + N$	13. $ax = x^3 + bx^2 + N$
6. $N = x^3 + ax$	14. $x^3 + N = bx^2 + ax$
7. $x^3 = bx^2 + N$	15. $x^3 + ax = bx^2 + N$
8. $bx^2 = x^3 + N$	16. $x^3 + bx^2 = ax + N$

[2] 1545 and 1663 have *latus cubicum*; 1570 has *latus cubi cum*.

It is evident, however, that of these the second, fifth, eighth, 11th, 13th, and 14th, by their very nature, have two solutions [and thus] differ from all [the rest] and depend on different rules. Hence by counting these cases twice, there are 22 primitive composite equations.[3] To each of these belong two derivative equations, one following the nature of the square, the other of the cube. For, although the derivatives are [really] infinite for any one case, they can all be reduced to one or the other of these two forms for purposes of discussing those for which a general rule can be given. Hence it is evident that there are exactly 44[4] of these, [5]for although the case of the first power equal to the square and the constant has two different solutions, I ought not, on that account, to say that there are two of them, for these two solutions follow from one and the same rule. Similarly, although the case of the cube and first power equal to the square and constant has three true solutions, yet this [statement] is not applicable to it either.[5] But it is not important to discuss the number of them. However, you should know that all primitive equations have two derivatives of diverse origin and that the primitive cases [from which they are] composed cannot be reduced to less than 18. Therefore, even contracting the number, there are at least 36 derivatives in any event, for the [solutions in the] cases of the first power equal to the constant and cube and of the square equal to the constant and cube are necessarily double, for it is evident how one solution differs from the other.

Given, therefore, an equation composed of three or four terms, if there is no number, first depress all powers by the lowest of them, so that the lowest turns into a number; then take the [new] lowest power and see if there is a rule for three powers that fits it. If the lowest is the square or cube root of the greatest, or if the square root of the smallest is the cube root of the greatest, then pursue the solution through the case similar to it among the 16. You then take such a root of this solution as is the [original] lowest power compared to the [new] lowest power. For the sake of ease, I placed all the derivatives opposite their originals in Chapter II — one of each order [next to] the other — including those composed of four terms which you have so carefully examined. The lowest power — that is, the least great

[3] 1570 and 1663 add, "And since the 15th has three solutions, there will be 24 [primitives]."

[4] 1570 and 1663 have 48.

[5] 1570 and 1663 omit this material.

after the number — is always the square root of one and the cube root of the other terms of this same case.

For example, if someone says,

1. $$x^4 + 2x^2 = 10,$$

you see that the original of this is the square and first power equal to the constant. Therefore seek the solution for

$$x^2 + 2x = 10,$$

which is $\sqrt{11} - 1$, and since x is the square root of x^2, the solution [for the original] will be $\sqrt{\sqrt{11} - 1}$.

2. $$x^6 + 2x^3 = 10.$$

The original of this is also the square and first power equal to the constant. Therefore, when

$$x^2 + 2x = 10,$$

the value of x is $\sqrt{11} - 1$, and since x is $\sqrt[3]{x^3}$ — namely the lowest power of the lowest [power] — the sought-for solution will be $\sqrt[3]{\sqrt{11} - 1}$.

3. $$(x^5)^2 + 2x^5 = 10.$$

You see that x^5 is $\sqrt{(x^5)^2}$. Say, therefore, that this is a derivative in the nature of the square. If, therefore,

$$x^2 + 2x = 10,$$

the value of x is $\sqrt{11} - 1$. Therefore, since x equals $\sqrt[5]{x^5}$, we say that the solution sought for is $\sqrt[5]{\sqrt{11} - 1}$.

4. $$x^6 + 3x^4 = 20.$$

Now you see that the original of this is the cube and square equal to the constant. When, therefore,

$$x^3 + 3x^2 = 20$$ [6]

x equals 2, and since x^2 is $\sqrt{x^4}$, the solution for x will be $\sqrt{2}$.

5. $$x^6 + 3x^4 + 10 = 15x^2.$$

You can see that the original of this in the table, or by the aforesaid reasoning, is the cube, square, and constant equal to the first power.

[6] 1663 has 10.

Therefore, find the solution for

$$x^3 + 3x^2 + 10 = 15x$$

which is 2, and since x is $\sqrt{x^2}$, I say that the solution will be $\sqrt{2}$.

6. $$x^9 + 3x^6 + 10 = 15x^3.$$

You say, as before, that the primitive of this one is the cube, square, and constant equal to the first power. Therefore, if

$$x^3 + 3x^2 + 10 = 15x,$$

x equals 2 and since x is $\sqrt[3]{x^3}$ we agree that the solution will be $\sqrt[3]{2}$. And since the primitive has two solutions, as has been noted, the derivative has the same number and the cube roots of the primitive's in this case or its square roots in the preceding case will satisfy it. This is generally true for all derivatives: they have as many solutions as their originals.

7. Again, let

$$x^9 = 3x^6 + 16.$$

Now, since multiplying $\sqrt{x^6}$, which is x^3, by x^6 produces x^9, therefore this [i.e., the sixth power] will be x^2[7] in a general primitive[8] equation, and the primitive will be the cube equal to the square and constant. If, therefore,

$$x^3 = 3x^2 + 16,$$

x will be 4. Since, then, x^2 is the lower power in the second equation and is the cube root of x^6, therefore, I say, [the first] will have $\sqrt[3]{4}$ as a solution. And so it goes for the others.

8.[9] And you may say the same for the ninth power plus the cube, for this is referable to the cube plus the first power, for as x is $\sqrt[3]{x^3}$, so x^3 is $\sqrt[3]{x^9}$. It may also be referred to the sixth and[10] second powers, for either multiplied by its own square root produces the same power, for from x^2 times x comes x^3 and from x^6 times x^3 comes x^9, but the former method is the easier.

[7] The text has x.

[8] The text has "derivative."

[9] 1663 omits this paragraph number.

[10] 1570 and 1663 omit this word.

CHAPTER XXV

On Imperfect and Particular Rules

The [foregoing] rules are called general for two reasons. The first is because the method itself is general, even though it is repugnant to the nature of a solution that it should be universal. Thus, if someone should say that the product of every number that is multiplied by itself is a square, that is a general rule. But it does not follow from this rule that I will know the square of every number, since it is impossible to know every number that is produced by multiplying another number by itself. A rule may also be called general because it exhausts a universal type of solution, although a solution does not exhaust the rule. Nevertheless there are also particular rules [and they are so called] since we cannot solve every given problem by them.

1. When the cube is equal to the first power and constant, divide the coefficient of x into two parts the product of one of which and the square root of the other is the constant of the equation. Then add one-fourth of that part the square root of which was taken to the other part. The square root of the whole added to one-half the square root of the part the root of which was used is the value of x.[1]

Example:

$$x^3 = 20x + 32.$$

Then 16 times $\sqrt{4}$ equals 32. Therefore add one-fourth of 4 to 16, making 17, the square root of which, plus 1, is the value of x. Hence x equals $\sqrt{17} + 1$.

2. When the cube is equal to the first power and constant, find two numbers whose product is the constant and one of which is the square root of the sum of the other and the coefficient of x. The square root is the value of x.[2,3]

[1] I.e., if $x^3 = ax + N$, let $a = f + g$ and let $f\sqrt{g} = N$, then $x = \sqrt{f + \frac{1}{4}g} + \frac{1}{2}\sqrt{g}$.

[2] I.e., if $x^3 = ax + N$, let $fg = N$ and let $f = \sqrt{a + g}$; then $x = \sqrt{a + g}$.

For example,

$$x^3 + 24 = 32x.$$

The two numbers producing 24 are 6 and 4, of which 6 is the square root of the sum of 32, the coefficient of x, and 4, the other number which was produced — for 6 is $\sqrt{36}$ — and therefore x equals 6.

3. When the cube is equal to the first power and constant, divide the coefficient of x into two such parts that the sum of each multiplied by the square root of the other is half the constant of the equation. The roots of these two parts added together constitute the value of x.[4,5]

For example,

$$x^3 = 10x + 24.$$

Ten divides into two parts, 9 and 1, either of which multiplied by the square root of the other makes 9 and 3, the sum of which is 12, one-half of 24. Therefore $\sqrt{9} + \sqrt{1}$, which are 3 and 1, added together, produce 4, the value of x.

4. When the cube equals the first power and constant, divide the coefficient of x into three parts in such proportion that the middle one multiplied by the sum of the roots of the first and third — or the third multiplied by the square root of the first plus the first times the square root of the third, which is the same thing — equals the constant. Then the sum of the aforesaid roots is the value of x.[6,7]

---

[3] 1570 and 1663 add:

But if one number is the square root of [the sum of] the coefficient of x and the other number [*et partis producentis*], the number therefore (if it is x) will, when multiplied by its square, produce the cube and when multiplied by the coefficient of x will produce the x's and when multiplied by the other number will, by supposition, produce the constant. Hence the cube will be equal to these x's plus the constant.

[4] I.e., if $x^3 = ax + N$, let $a = f + g$ and let $f\sqrt{g} + g\sqrt{f} = N/2$; then $x = \sqrt{f} + \sqrt{g}$.

[5] 1570 and 1663 add:

[This is so] because the sum of the cubes of two numbers and of their mutual parallelepipeds is to the remaining four parallelepipeds as the sum of their squares is to twice their product [i.e., $(f^3 + g^3 + f^2g + fg^2):(2f^2g + 2fg^2) = (f^2 + g^2):2fg$]. But by supposition, the square roots of these parts of the coefficient of x (which coefficient is equal to the sum of the squares [of these square roots]) produce, when multiplied by their mutual squares, half the constant. From these, therefore, the coefficient is produced. Therefore the sum of these roots is x.

[6] I.e., if $x^3 = ax + N$, let $a = f + g + h$, and let $f:g = g:h$, and let $g(\sqrt{f} + \sqrt{h}) = N$ or let $h\sqrt{f} + f\sqrt{h} = N$; then $x = \sqrt{f} + \sqrt{h}$.

[7] 1570 and 1663 add:

Since the ratio of the squares of the parts plus their mean surface is to the mean surface as the sum of the cubes and four of their parallelepipeds is to the two remaining

For example,

$$x^3 = 19x + 30.$$

Now 19 makes three proportional parts — 9, 6 and 4 — of which the second, which is 6, multiplied by 5, the sum of the roots of the first and third, makes 30. Therefore 5, the sum of the roots, is the value of x.

5. When the cube is equal to the first power and constant, look for two numbers the sum of which multiplied by their product produces one-third the constant. The squares of these are equal to the sum of the coefficient of x and the product of one multiplied by the other. Therefore the sum of the numbers is the solution for x.[8,9]

---

parallelepipeds [i.e., $(f^2 + h^2 + fh):fh = (f^3 + h^3 + 2f^2h + 2fh^2):(f^2h + fh^2)$] and the two squares have the surface as a mean proportional [i.e., $f^2:fh = fh:h^2$], therefore divide the cube by the [sum of the] root[s] of the squares in accordance with the ratio of their bases. If the mutual product [of these roots] or the mean surface times x produces the constant [i.e., if $(f + h)fh = N$], then x times the remaining three parts of the base [i.e., $(f + h)(f^2 + fh + h^2)$] will give the six remaining bodies of the cube [i.e., $f^3 + 2f^2h + 2fh^2 + h^3$]. Hence the cube is equal to the first power and the constant.

The correctness of the midpart of this is quite uncertain, so I give the original for the benefit of anyone who prefers another rendering: *Quia proportio quadratorum partium cum superficie media ad mediam superficiem est sicut aggregati cuborum cum quatuor parallelipedis ad duo reliqua parallelipeda: et illa duo quadrata habent superficiem in media proportione, igitur diviso cubo iuxta rationem basis latere quadratorum, si producant numerum invicem mutuo ducta seu media sit perficies* [so in 1570; 1663 has *sit superficies*, probably the result of correcting the *sit* of 1570 to *su* and forgetting to delete the *sit* itself] *in rem, ex re in reliquas tres partes basis, fient sex corpora residua cubi: ergo cubus ille est aequalis rebus et numero.*

[8] I.e., if $x^3 = ax + N$ and $(f + g)fg = N/3$, then $f^2 + g^2 = a + fg$, and $x = f + g$.

[9] 1570 and 1663 add:

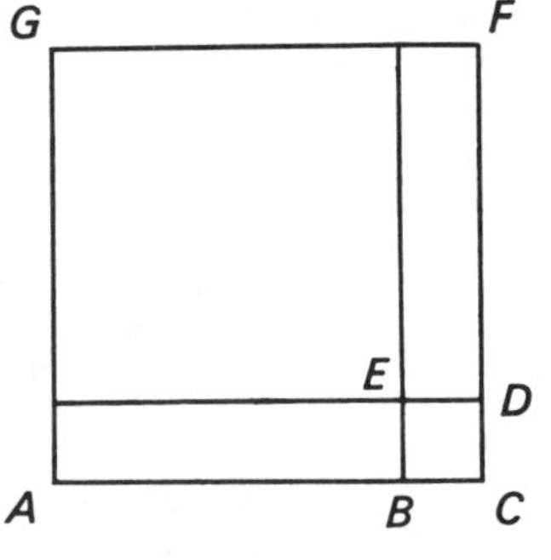

This note is the converse of the general rule. Since AE [1663 has AC] plus the square of the difference [between AB and BC] is equal to the coefficient of x, therefore, by the demonstration in the book on proportions, this number of x's will equal the cubes of AB and BC. [I.e., if $x^3 = ax + N$, and if $N = 3(f + g)fg$, then $fg + (f - g)^2 = a$, and therefore $[fg + (f - g)^2]x = f^3 + g^3$. It can also be demonstrated that the product of AB and BC^2 and of BC and AB^2 is equal to the product of AC and AE [i.e., $fg^2 + f^2g = (f + g)fg$]. The latter, however, has heretofore been supposed to equal one-third the constant [i.e., $(f + g)fg = N/3$]. Therefore three times this [is equal to] three times that and $AC^3 = N + ax$, as was proposed [i.e., $(f + g)^3 = 3(f + g)fg + [fg + (f - g)^2](f + g)$].

For example,

$$x^3 = 7x + 90.$$

Now 3 and 2, multiplied together, produce 6 which, when multiplied by 5, their sum, makes 30, or one-third of 90. The difference between 13, the sum of their squares, and 6, their product, is 7, the coefficient of x. Therefore 5, their sum, is the value of x.

6. When the cube is equal to the first power and constant, find a cubic number the cube root of which times the coefficient of x produces either the sum of the cubic number and the constant or their difference. Then either x plus the same cube root will be a common divisor for x^3 plus this same cubic number and for the coefficient of x plus the sum of the constant and the cubic number, or x minus this same cube root will be a common divisor for x^3 minus the cubic number and for the coefficient of x minus the difference between the constant of the equation and the cubic number. Thus you arrive at the value of x.

Example:

$$x^3 = 16x + 21.$$

Now since adding 27, a cubic number, to 21 makes 48, which is the product of 3, the cube root of 27, and 16, the coefficient of x, therefore I say that $x + 3$ will be a common divisor, if 27 is added to both sides [of the equation] — i.e., to x^3 and to $16x + 21$. Then, having made the division, you will have

$$x^2 - 3x + 9 = 16.$$

Therefore

$$x^2 = 3x + 7,$$

and x equals $\sqrt{9\frac{1}{4}} + 1\frac{1}{2}$.

And similarly, if we say

$$x^3 = 4x + 15.$$

In this case, subtract 15 from 27, a cubic number, leaving 12, which contains 4, the coefficient of x, multiplied by 3, the cube root of 27. Therefore I say that, if 27 is subtracted from both sides, it will make

$$x^3 - 27 = 4x - 12.$$

Then you divide both by $x - 3$, a common divisor, and this makes

$$x^2 + 3x + 9 = 4.$$

[But] for this no solution follows — although you have [already] arrived at the means of equating in the subtraction — unless, perhaps, it be something by way of a negative hodge-podge.[10]

7. When the cube is equal to the first power and constant, subtract three-fourths of the square of x from the coefficient of x and add or subtract the square root of the remainder to or from one-half of x. The sum multiplied by the square of the difference and the difference multiplied by the square of the sum [when added together] produce the constant of the equation.[11]

Example:

$$x^3 = 14x + 8,$$

and x is 4. The square of this is 16, three-fourths of which is 12. Subtracted from 14, the coefficient of x, the remainder is 2, the square root of which add to and subtract from 2, one-half of 4, the value of x. These make $2 + \sqrt{2}$ and $2 - \sqrt{2}$. Therefore, I say that by mutually multiplying one by the square of the other [and adding the products] 8, the constant of the equation, is produced.

8. When the cube is equal to the first power and constant, divide one-half the constant of the equation by the value of x and add the quotient to the coefficient of x. From this sum subtract three-fourths the square of the same x. Add or subtract the square root of the remainder to or from one-half the solution for x, and the parts resulting when mutually multiplied, one by the square of the other, give one-half the constant of the equation.[12]

For example,

$$x^3 = 14x + 8,$$

and x is 4. Divide 4, one-half of 8, by 4, the value of x. The result is 1. Add this to 14, making 15. Subtract 12, three-fourths the square of x, leaving 3, the square root of which add to and subtract from 2, one-half

[10] *syncerum,* here taken to be a syncopated form of *syncerastum.* But perhaps it should be taken as equivalent to *sincerum,* in which case the translation would be "something in the shape of a pure negative."

[11] I.e., if $x^3 = ax + N$,

$$N = (\tfrac{1}{2}x + \sqrt{a - \tfrac{3}{4}x^2})(\tfrac{1}{2}x - \sqrt{a - \tfrac{3}{4}x^2})^2 + (\tfrac{1}{2}x - \sqrt{a - \tfrac{3}{4}x^2})(\tfrac{1}{2}x + \sqrt{a - \tfrac{3}{4}x^2})^2.$$

[12] I.e., if $x^3 = ax + N$,

$$\frac{N}{2} = \left(\frac{x}{2} + \sqrt{\frac{N}{2x} + a - \tfrac{3}{4}x^2}\right)\left(\frac{x}{2} - \sqrt{\frac{N}{2x} + a - \tfrac{3}{4}x^2}\right)^2 + \left(\frac{x}{2} - \sqrt{\frac{N}{2x} + a - \tfrac{3}{4}x^2}\right)\left(\frac{x}{2} + \sqrt{\frac{N}{2x} + a - \tfrac{3}{4}x^2}\right)^2.$$

of x, and you will have $2 + \sqrt{3}$ and $2 - \sqrt{3}$. By mutually multiplying one of these by the square of the other [and adding the products] 4 is produced, which is one-half the constant.

9. When the first power is equal to the cube and number, find a number which, multiplied by the square root of the sum of itself and the coefficient of x, produces the constant of the equation. Then one-half that root plus or minus the root of the difference between the coefficient of x[13] and three-fourths the same sum is the value of x.[14]

Example:

$$x^3 + 12 = 34x.$$

Then, since adding 2 to 34 [gives 36 and] the product of this same 2 and 6, the square root of 36, the sum of 2 and 34, is 12, the constant of the equation, I say that if to or from 3, one-half the root of 36, is added or subtracted $\sqrt{7}$, the [square root of the] difference between 34, the coefficient of x,[13] and 27, which is three-fourths the square of 6 or of the sum [given above], what arises is the value of x: $3 + \sqrt{7}$, or $3 - \sqrt{7}$.

10. When the first power is equal to the cube and constant, subtract the constant from such a number that the cube root of the difference times the coefficient of x produces the number.[15] Then x minus the cube root of the difference will be a common divisor, [the number] having been subtracted [from both sides of the equation]. This rule is similar to the sixth, as the preceding one was to the second.

Example:

$$16x = x^3 + 21.$$

Subtract [21] from 48, leaving 27, the cube root of which is 3. This multiplied by 16, the coefficient of x, produces 48. Therefore subtract 48 from both sides of the equation, making

$$x^3 - 27 = 16x - 48.$$

The common divisor will be $x - 3$, and this results in

$$x^2 + 3x + 9 = 16.$$[16]

[13] In both these places, the text has "the constant."

[14] I.e., if $x^3 + N = ax$, let $y\sqrt{a + y} = N$; then $x = \frac{\sqrt{a + y}}{2} \pm \sqrt{a - \frac{3}{4}(a + y)}$.

[15] The text has "subtract from the constant such a number that the cube root of the difference times the coefficient of x produces the number subtracted."

[16] 1570 and 1663 have 19.

Therefore

$$x^2 + 3x = 7$$

and x will be $\sqrt{9\frac{1}{4}} - 1\frac{1}{2}$.

11. When the first power is equal to the cube and constant, divide the coefficient of x into three proportional parts the second of which multiplied by the difference between the roots of the first and third — i.e., the difference between the product of the first of which and the root of the third and the product of the third and the root of the first[17] — equals one-third the constant of the equation. The difference between these roots will be the value of x.[18] This is similar to the fourth.

Example:

$$19x = x^3 + 18.$$

Divide 19 into three proportional parts — 4, 6 and 9 — of which the middle, 6, multiplied by the difference between the roots of 9 and 4, which is 1, makes 6, one-third of 18, the constant. Therefore I say that 1, the difference between these roots is the value of x.

12. When the first power equals the cube and constant, divide the coefficient of x by the cube root of the constant of the equation and make two parts from the result, one of which multiplied by the square of the other gives the constant of the equation. Then the quantity proportional between the cube[19] root of the constant and the part which you multiply by the square of the other in order to derive the constant, is the value of x.[20]

Example:

$$18x = x^3 + 8.$$

I divide 18 by 2, the cube root of 8; 9 results, the two parts of which are 8 and 1, one of which — which is 8 — multiplied by the square of the other — which is 1 — makes 8, the constant of the equation. Therefore 4, the number proportional between 8, the part of 9 which you multiplied by the square of 1, the other part, and 2, the cube root of 8, the constant of the equation, is the value of x.

[17] The formula between the dashes is the same as the whole formula ahead of it; it is not a difference to be multiplied by the second proportional.

[18] I.e., if $x^3 + N = ax$, let $a = f + g + h$ and let $f:g = g:h$; then $g(\sqrt{f} - \sqrt{h}) = N/3$, and $x = \sqrt{f} - \sqrt{h}$.

[19] 1545 has *rubicam*; 1570 and 1663 have *cubicam*.

[20] I.e., if $x^3 + N = ax$, let $\frac{a}{\sqrt[3]{N}} = f + g$, and let $fg^2 = N$. Then $\sqrt[3]{N}:x = x:f$.

13. When the cube and constant are equal to the first power, divide one-third the coefficient of x into two parts which, multiplied by their roots, produce two numbers the sum of which is equal to one-half the constant of the equation. The sum of these roots is the value of x.[21] This is similar to the third rule.

Example:

$$15x = x^3 + 18.$$

I take 5, one-third of 15, and from this I make two parts, 4 and 1, which multiplied by their roots, 2 and 1, produce 8 and 1, the sum of which, 9, is one-half of 18, the constant of the equation. Therefore I say that 3, the sum of these roots, is the value of x.

Now you also know from the general rule that, whenever the coefficient of x can be broken into two parts one of which multiplied by the root of the other produces the constant of the equation, that square root is the value of x and that this can be in two forms, and you also know how it resolves itself into a *binomium* or *recisum* and whole numbers.[22] Hence, although these [rules] are similar to the first rule, yet since we have snatched this from the general rule as if by violence, it should be enough to be forewarned at this point.

14. When the number equals the cube and square, see whether you can divide the coefficient of x^2 into two parts the product of one of which and the square of the other makes the constant of the equation. Then multiply the one that was not squared by that which was squared plus one-fourth of that which was not squared. The square root of the product minus one-half of the part which was not squared is the value of x.[23]

Example:

$$x^3 + 20x^2 = 72.$$

Divide 20 into two parts, 18 and 2. One of these multiplied by the square of the other, makes 72, for 18×4 equals 72. I say that if 18 is multiplied by $6\frac{1}{2}$, the sum of 2, the second part, and $4\frac{1}{2}$, one-fourth of 18, it produces 117, the square root of which minus 9, one-half of 18, gives the value of x as $\sqrt{117} - 9$.

15. When the square is equal to the cube and constant, look for a number — not less than one-fourth the coefficient of x^2 nor more than

[21] I.e., if $x^3 + N = ax$, let $\frac{1}{3}a = f + g$ and let $f\sqrt{f} + g\sqrt{g} = N/2$. Then $x = \sqrt{f} + \sqrt{g}$.

[22] *in binomio vel reciso et integris.*

[23] I.e., if $x^3 + ax^2 = N$, let $a = f + g$ and let $fg^2 = N$; then $x = \sqrt{f(g + f/4)} - f/2$.

one-third — with which, by dividing the constant of the equation, you arrive at a square number one-half the square root of which added to the coefficient of x^2 [24] makes four times the divisor. Then the value of x is twice the divisor plus or minus the square root of the product[25] of four times the divisor and the difference between the coefficient of x^2 [26] and three times this same divisor.

Example:

$$x^3 + 48 = 10x^2.$$

Now since 3 — which is not less than one-fourth of 10, the coefficient of x^2, nor greater than one-third the same — divided into 48 produces 16, one-half the square root of which, 2, added to 10, the coefficient of x^2, makes 12, four times the divisor, or 3, therefore I say that if to or from twice the divisor, which is 6, is added or subtracted the square root of the product of 12,[27] four times 3, the divisor, and 1, the difference between 10, the coefficient of x^2,[26] and 9, three times 3, the divisor, which product is also 12, we will have constituted both solutions: $6 + \sqrt{12}$, and $6 - \sqrt{12}$.

Note. You now understand from these cases that rules and problems far beyond them may be found with ease which could scarcely be solved before. These rules, truly, are founded on the demonstrations of the sixth chapter, and I need not set them out here, since they are self-evident to those familiar with our books on Euclid, and he who does not know these will neither care nor seek for them since they are no friends of his.[28]

16. It is worth noting that none of these rules can be [considered] general with respect to a solution. The method used [to show this] in one case will suffice to show it in the others. Let us take, therefore, the case nearest at hand, which, since [it admits] of a multiple solution, will make this the more readily believable. Now let

$$x^3 + N = 7x^2,$$ [29]

and let $2\frac{2}{3}$ be an assumed number — that is, the [sort of] number which was discussed in the second rule of the sixth chapter. According to

[24] 1545 and 1570 have *numero quadratorum*; 1663 has *numero quadratae.*

[25] *tunc aestimatio rei est duplum numeri divisoris p: vel m: radice producti.*

[26] The text has x in both of these places.

[27] 1663 omits "of 12."

[28] *quoniam non sunt ei necessariae.*

[29] 1663 omits the coefficient 7.

that rule, then, the value of x will be $\sqrt{16} + 2\frac{2}{3}$ or $6\frac{2}{3}$. The difference [between this and] the coefficient of x^2 is $\frac{1}{3}$ and, from the demonstration given at the beginning of the third book, the product of $6\frac{2}{3}$ and the square of $\frac{1}{3}$ is a fraction and is $\frac{20}{27}$. Contrariwise, the product of $\frac{1}{3}$ and the square of $6\frac{2}{3}$ is also a fraction, namely $14\frac{22}{27}$. Wherefore, having assumed the coefficient of x^2 to be an integer and the solution to be made up of fractional numbers, the constant of the equation, which is the amount by which the parts of the squares which are rational exceed the cube, can never be a whole number, but the constant of such an equation is the product of one part of the coefficient of x and the square of the other.

Having shown this, I take

$$x^3 + N = 7x^2.$$

It is evident, from the demonstrations of our seventh book on Euclid and from the rules of the sixth book on raising a number to its square or cube, that the greatest product of one part of 7 and the square of the other is $50\frac{22}{27}$.[30] Therefore 7 can be divided so that multiplying one part by the square of the other produces [various] whole numbers between one and 50. [The parts, of course, have to be] integers, not fractions, according to the demonstrations. But no divisions [of 7] into integers, unless it be a three-fold division, can produce anything except 6, 20, 36, 48, and 50,[31] as is shown in the figure. Therefore, the other 45 numbers can by no means be exhausted in a solution of this type. [The rule], therefore, is particular[32] and, indeed, exceedingly particular.[32] Yet you should not believe that in other cases the constant cannot be taken care of through the other part of a *binomium* or *recisum*, as we have frequently shown in the examples.

7			
1	6	36	6
2	5	50	20
3	4	48	36

17. When the cube and constant are equal to the first power, divide the square root of the coefficient of x into two parts, the first of which multiplied by twice the square of the second and the second by the square of the first gives the constant of the equation. In this case, the second part will be the value of x.[33]

[30] I.e., $2\frac{1}{3} \times (4\frac{2}{3})^2$.

[31] 1570 and 1663 have 30.

[32] 1545 has *particularis*; 1570 and 1663 have *specialis*.

[33] I.e., if $x^3 + N = ax$, and if $\sqrt{a} = f + g$, and if $2fg^2 + f^2g = N$, then $x = g$.

Example:

$$x^3 + 48 = 25x.$$

Now since 5, the square root of[34] 25, divides into two parts, 3 and 2, by the multiplication of which — [i.e.,] 2 by 18, the square of 3 doubled and 3 by 4, the square of 2 — 48 results, therefore I say that 3, the part whose square is doubled, is the value of x.

18. When the cube and square equal the constant, and the constant of the equation is equal to the difference between two numbers[35] whose product is the same as the cube of one-third the coefficient of x^2 times the difference between the cube roots of these numbers, then the difference between these cube roots is the value of x,[36] as is shown in the margin here. Hence x is easy [to find].[37]

$x^3 + 22\frac{1}{2}x^2$		$= 98$
3375	125	27
	98	
$7\frac{1}{2}$	5 — 3	
		2
$421\frac{7}{8}$		8
3375		

[34] 1570 and 1663 have & instead of ℞.

[35] *et duo numeri differentes in numero aequationis.*

[36] I.e., if $x^3 + ax^2 = N$, and $N = f - g$, and $fg = \left(\frac{a}{3}\right)^3 (\sqrt[3]{f} - \sqrt[3]{g})$, then $x = \sqrt[3]{f} - \sqrt[3]{g}$.

[37] 1570 and 1663 add at this point:

Corollary. From this appears one remarkable thing: viz., that in those cases in which the given constant is a composite, the solution can frequently be discovered very easily, but if it is a prime quite rarely. Since it does not happen that two proportional or fractional parts of a whole number [multiplied] together produce a whole number, how much less [is this likely to be] so [if one is multiplied] by the square or root of the other. This has been presupposed in many of these rules.

Corollary. Since in accordance with the 14th rule of this [chapter], two such parts can be made of $22\frac{1}{2}$, the coefficient of x, that one of them times the square of the other makes 98, the constant of the equation, and the solution is the difference between the square root of the product of one of them and one-fourth itself and the remainder of one-half the same part, therefore assuming the first part to be y, we multiply this by $22\frac{1}{2} - \frac{3}{4}y^2$, the square root of which is 2 greater than $\frac{1}{2}y$. Therefore $\frac{1}{2}y + 2$ is equal to this root. Hence the first part is $10\frac{1}{4} + \sqrt{101\frac{1}{16}}$, the other $12\frac{1}{4} - \sqrt{101\frac{1}{16}}$.

CHAPTER XXVI

Shows Certain Greater Rules That Are Entirely Particular

1. If the fourth and first powers are equal to the square and constant and if, having divided the coefficient of x and the constant by the coefficient of x^2, half the result of dividing the coefficient of x is [equal to] the square root of that which came out of the division of the constant of the equation, then take the square root of the constant of the original equation, add to it one-fourth the coefficient of x^2, take the square root of the sum, from this subtract the square root of the same one-fourth the coefficient of x^2, and the remainder is the value of x.[1]

PROBLEM

For example: Four men form an organization. The first deposits a given quantity, the second deposits the fourth power of one-tenth the first, the third five times the square of one-tenth the first, and the fourth 5. Let the sum of the first and second equal the sum of the third and fourth. The problem is to find how much each deposited.

Assume that the first deposited $10x$. Therefore the second's deposit will be x^4, the third's $5x^2$. The fourth, moreover, as was said, deposited 5. Therefore

$$x^4 + 10x = 5x^2 + 5.$$

By dividing, then, the coefficient of x by the coefficient of x^2, 2 results, [the root of] one-half of which is $\sqrt{1}$. This [also] results from dividing 5, the constant of the equation, by the coefficient of x^2 which is 5. Therefore take the square root of 5, the constant of the equation, and add to it one-fourth the coefficient of x^2, making $\sqrt{5} + 1\frac{1}{4}$. Take the square root of this whole sum, which is $\sqrt{\sqrt{5} + 1\frac{1}{4}}$,[2] and from this

[1] I.e., if $x^4 + ax = bx^2 + N$, and if $a/2b = \sqrt{N/b}$, then $x = \sqrt{\sqrt{N} + b/4} - \sqrt{b/4}$.

[2] 1663 omits the universal radical sign.

subtract [the square root of] one-fourth the coefficient of x^2 and you will have as the value of x $\sqrt{\sqrt{5} + 1\frac{1}{4}} - \sqrt{1\frac{1}{4}}$. [The various amounts, then,] will be, as you see,

1st $\sqrt{\sqrt{50{,}000} + 125} - \sqrt{125}$

2d $17\frac{1}{2} + \sqrt{500} - \sqrt{\sqrt{612{,}500} + 781\frac{1}{4}}$

3d $12\frac{1}{2} + \sqrt{125} - \sqrt{\sqrt{78{,}125}[3] + 156\frac{1}{4}}$

4th 5

2. Likewise, if the fourth power is equal to the square, first power, and constant, a similar rule will hold under the same conditions, and the method of solution will be the same except that, at the end, we add the square root of one-fourth the coefficient of x^2 to the square root of the whole, whereas in the preceding rule we subtracted it. For example, if

$$x^4 = 5x^2 + 10x + 5,$$

x will equal $\sqrt{\sqrt{5} + 1\frac{1}{4}} + \sqrt{1\frac{1}{4}}$.

And the reason for this in these rules is that the square root of x^4 is x^2, and the square root of $5x^2 - 10x + 5$ is $\sqrt{5}$[4] $- \sqrt{5x^2}$ or $-x\sqrt{5}$. Therefore,

$$x^2 + x\sqrt{5} = \sqrt{5}$$

and the solution is known and is the same as when

$$x^4 + 10x = 5x^2 + 5.$$

By the same reasoning, if

$$x^4 = 5x^2 + 10x + 5,$$

x^2 will equal $x\sqrt{5} + \sqrt{5}$, wherefore x is known.

3. When the fourth power, square, and[5] first power are equal to the cube and constant and the constant is 2 more than the coefficient of x^2 and the coefficients of x and x^3 are the same and one-half the coefficient of x is the square root of the constant, then take the square of one-fourth the coefficient of x, add 1 to it, subtract the square root of the sum of 1 and the square of one-half the coefficient of x, and add

[3] The text omits the internal radical sign.

[4] 1663 omits the 5 under the radical.

[5] 1545 and 1570 have *est*; 1663 has *et*.

or subtract the square root of the remainder to or from one-fourth the coefficient of x. The result is the value of x.[6]

Example,

$$x^4 + 34x^2 + 12x = 12x^3 + 36.$$

Now you notice that the coefficient of x is the same as the coefficient of x^3 and that one-half the coefficient of x is the square root of 36, the constant, and that the constant is 2 more than the coefficient of x^2. Therefore multiply 3, one-fourth of 12, the coefficient of x, by itself, making 9. Add 1, according to the rule, making 10. Subtract $\sqrt{37}$, the sum of the square of one-half the coefficient of x and 1, from this, making $10 - \sqrt{37}$.[7] Add or subtract the square root of this entire quantity to or from 3, one-fourth the coefficient of x, and you will have $3 + \sqrt{10 - \sqrt{37}}$, and $3 - \sqrt{10 - \sqrt{37}}$, as the values of x.

4. Now the method of arriving at such rules as these comes from the great rule, whence also we give this chapter its name. It is that you [first] solve any problem simply and then by a great rule or some other one. Then you observe the necessary conditions for the transition from one to the other, afterwards noticing how you arrived at the value of x, and you then construct a new rule for the unknown case by this means.

Example: Divide 6 into two parts such that the cube of the smaller, the square of the greater, and eight times this same greater are all in proportion. I say that by the great rule you come to this: that the proportion of these parts will be $\sqrt[3]{8}$ or 2. Hence we divide 6 by $\sqrt[3]{8} + 1$, and the value of x will result.

But by using the unknown,[8] we will have

$$x^4 + 24x^2 + 144 = 8x^3 + 96x.$$

We say, therefore, since the fourth power, second power, and constant are equal to the cube and first power, that we have to find a number which, multiplied by the constant of the equation, produces a number the square root of which multiplied by 6, in accordance with the rule, produces a number which, divided by the first number which you multiplied, produces the coefficient of x^2. Now if to this forementioned number which you multiplied by the constant of the equation you add 3, according to the rule, and if this is multiplied by the fourth

[6] I.e., if $x^4 + bx^2 + ax = cx^3 + N$, and if $N = b + 2$, and if $a = c$, and if $a/2 = \sqrt{N}$, then $x = a/4 \pm \sqrt{(a/4)^2 + 1 - \sqrt{(a/2)^2 + 1}}$.

[7] 1545 and 1570 have *10 m:* ℞ *37*; 1663 have *10 abiice* ℞ *37.*

[8] I.e., by starting with $6 = x + (6 - x)$.

root of the number which you produced in the beginning, a number will result which, divided by the first number, produces the coefficient of x^3. The coefficient of x multiplied by the first number will then be four times the cube of this fourth root. Then, I say, subtract 1, according to the rule, from the first number which you multiplied and take the cube root of the remainder and add 1 to it, according to the rule, and divide this fourth root by the sum, and what results is the value of x.[9] The reason for this is that in such a case the coefficient of x^4 derives from the multiplicand plus 1, the coefficient of x^3 from the dividend times the multiplicand plus 4, the coefficient of x^2 from six times the square of the dividend, the coefficient of x from four times the cube of the dividend, and the constant of the equation is the fourth power of the dividend.[10] In this problem, moreover, I call 6 the dividend and 8 the multiplicand.

For example,

$$x^4 + 6x^2 + 4 = 3\tfrac{1}{2}x^3 + 8x.^{11}$$

Let the first number be y^2 and multiply by 4, making $4y^2$. The square root of this is $2y$. Multiply this by 6, according to the rule, making $12y$, and divide this by y^2. What results equals 6. Therefore, $6y^2$ equals $12y$, wherefore y equals 2. We, however, used y^2 instead of y. Therefore, the first number, or the multiplicand, will be 4, and when the other conditions which were stated are fulfilled, 2 will be the dividend. Having divided this by $\sqrt[3]{3} + 1$, the value of x will appear. Of this we spoke in the sixth chapter.

[9] I.e., if $x^4 + bx^2 + N = cx^3 + ax$, then $b = \dfrac{6\sqrt{Ny}}{y}$; $c = \dfrac{(y + 3)\sqrt[4]{Ny}}{y}$; $ay = 4(\sqrt[4]{Ny})^3$; and $x = \dfrac{\sqrt[4]{Ny}}{\sqrt[3]{y - 1} + 1}$.

[10] These statements are all true, given x and $(6 - x)$ as the starting point; the equation given at the beginning of this paragraph represents the same after dividing through by 9.

[11] 1570 and 1663 have $7x$.

CHAPTER XXVII

On Transition from One Particular Equation to Another

1. Transition is made from one particular equation to another after this fashion:

$$x^3 + 2x^2 + 56 = 41x,$$

for which one solution is $3 + \sqrt{2}$. I am seeking, with the same solution, [the unknown terms in]

$$y^3 + 7y^2 + N = ay.$$

Multiply the difference between the coefficients of the squares, which is 5,[1] by twice the part of the [given] solution which is a whole number — that is, by 6 — making 30, to which add 41, the coefficient of x, making 71, the coefficient of y. Then multiply each part of the solution by itself, making 2 and 9, the difference between which is 7. Multiply this by 5, the difference between the coefficients of the squares, making 35, which add to 56, since 3 is greater than $\sqrt{2}$, making the constant of the [second] equation 91. Therefore,

$$y^3 + 7y^2 + 91 = 71y,$$

the given solution being $3 + \sqrt{2}$. If the root were greater than the number, you would subtract 35 from 56, and 21 would be the [second] constant.

2. I may also say that it is not permissible to go from one equation to another if the nature of the terms remains the same and if the value of x is both the same and irrational — i.e., neither a whole number nor a fraction. For example, let

$$x^3 + 3x = 10.$$

[1] 1570 and 1663 have 2.

X, then, equals $\sqrt[3]{\sqrt{26}+5} - \sqrt[3]{\sqrt{26}-5}$. I say that with this solution, x^3 plus x, whatever its coefficient, cannot equal any other number, even up to infinity. Let, for example,

$$x^3 + 9x = 18.$$

Since, then, the [value of] x is the same, namely the aforesaid cube roots, the cube will remain the same in both. Therefore, according to the third book,

$$x^{3}\,^{2} + 9x + 10 = x^3 + 3x + 18.$$

Hence, subtracting the common cube,

$$9x + 10 = 3x + 18.$$

Therefore, $6x$ equals 8, wherefore x equals $1\frac{1}{3}$, a rational number and not the cube roots that were given, which is contrary to the supposition.

3. Likewise, it is evident from the preceding that, if the solution remains the same, the cube with a greater coefficient and x with a greater coefficient cannot equal any [i.e., the same] constant, for dividing through by the coefficient of x^3 we would have, as before, a cube and first power equal to a constant which has already been shown to be impossible. The same reason works in every case, for if I should say,

$$x^3 = 6x + 2$$

or

$$x^4 = 6x + 2,$$

I say that, with the same solution, the cube or fourth power can equal no [other] rational [number of] x's plus a rational constant. I say "rational" in order not to bar [the possibility] that a solution might follow if either the coefficient of x or the constant were assumed to be irrational.

4. From this it follows that the rule also holds for different powers when the value of x is neither a rational number nor a simple root of the same nature as the middle power. For example,

$$2x^3 + 10 = x^4.$$

The value of x is neither a [rational] number nor the simple cube root

[2] The text has x^9.

of any rational number. I say that the fourth power, with such a solution, can equal no [other] cubes and number. This is evident, since, the change having been made and the fourth power eliminated, the cube would be left equal to a number. Therefore the value of x would necessarily be a number or the cube root of a number, which is contrary to the supposition.

CHAPTER XXVIII

On the Operation of Promic or Mixed Roots and Allela *Roots*

1. We have heretofore shown that there are three kinds of promic roots: The minor, when a square root is compared to the sum of its square and itself; this sum is called the minor promic. The mean, when a cube root is compared to the sum of itself and its cube; this sum is called the mean promic. But the major promic root occurs when the fourth root of a number is compared to the sum of itself and the number of which it is the fourth root; this sum is called the major promic, as in this example: The major promic of 3 is 84, and 3 is the major promic root of 84. Ordinary operations, however, do not deal with these since they are like anomalous words in grammar; they cannot be multiplied, divided, added or subtracted, but have their own proper function, which is called transition.

2. When, therefore, you multiply a minor promic by its promic root and add to the product the same promic, the square root of the whole will be the mean promic of the square root of the minor promic root. For example, I multiply 3, the minor promic root of 12, by 12, making 36. I add 12, the minor promic, to this, producing 48, the square root of which — and it is $\sqrt{48}$ — is the mean promic of $\sqrt{3}$, 3 having been the minor promic root of 12, for cubing $\sqrt{3}$ makes $\sqrt{27}$, and adding to this the same $\sqrt{3}$ makes $\sqrt{48}$. Therefore $\sqrt{3}$ is the mean promic root of $\sqrt{48}$, as was proposed.

3. When you multiply a mean promic by its promic root it produces the minor promic of the square of the mean promic root. Example: Multiply 3, the mean promic root of 30, by 30, making 90,[1] the minor promic of 9, the square of 3, which was the mean promic root of this same 30.

[1] 1570 has 902; 1663 has 900.

4. And when the major promic is squared and the product is divided by the square of the major promic root, the [ratio of the] quotient to the cube of this same promic root is the square [of the cube] plus twice [the cube] plus 1. For example: I take 18, a promic major, and square it, making 324. I divide this by 4, the square of 2, the promic major root of 18, and 81 results which, with respect to 8, the cube of 2, the same promic root, is its square plus twice itself plus 1.

5. Roots are called *allela* when, by the multiplication of each of two numbers by the square of the other, two numbers arise. Thus, when I take 2 and 3, these are called the *allela* roots of 12 and 18, for from 2 times 9 comes 18 and from 3 times 4 comes 12. These roots, however, are discoverable in another way: Multiply each [number] by itself and divide the product by the other and the cube roots of the quotients are *allela* roots. For example, I wish the *allela* roots of 4 and 8. Square 8, making 64; divide this by 4; 16 results. Again, square 4, making 16; divide this by 8; 2 results. Therefore the cube roots of 16 and 2 are the *allela* roots of 4 and 8. Thus the *allela* [roots] of 6 and 18 are $\sqrt[3]{54}$ and $\sqrt[3]{2}$.

Corollary. From this it is evident that all *allela* roots are the cube roots of numbers in a ratio to each other which is triplicate to that in which the numbers originally proposed stood and that these [the original numbers] are the mean proportionals.[2]

6. From this, therefore, the operations in these [cases] are evident, for when they are discovered they reduce to cube roots with which you can work back again. The operation having been completed, you reducc all to *allelae*.

[2] *Ex quo patet, quod omnes* ℞ *allellae, sunt* ℞ *cubicae numerorum, se habentium in triplicata proportione, in qua se habent sui solidi propositi priores, et hi sunt medii proportionales.*

CHAPTER XXIX

On the Rule of Method[1]

This rule of method is so called because it shows the method for constructing as many rules of commerce as you wish. It is very useful to teachers of arithmetic in teaching that art, certain easier [methods] having been discovered [by means of it]. With its help, we constructed the greater part of the sixth book. This, then, is the rule: Solve any given problem by any means you can, either [by using] an unknown or with the help of the sixth book. They lay aside the unknown and the other rules and use those operations which you can best use, looking always for brevity, and you will have the rule for the method of [solving] any similar problem.

For example: Seven feet of green silk and three of black cost 72 *denarii* and, at the same price, two of green and four of black cost 52 *denarii*. We wish to know their price. Assume that x is the price of one foot of the green silk. Then seven cost $7x$. Hence three feet of the black cost 72 *denarii* minus $7x$, and one costs one-third of this, namely 24 *denarii* minus $2\frac{1}{3}x$. So four feet of the black will cost 96 *denarii* minus $9\frac{1}{3}x$, and two of the green are worth, by supposition, $2x$. Therefore two feet of green and four of black are valued at 96 *denarii* minus $7\frac{1}{3}x$ and are also worth 52 *denarii*. Hence

$$96 \textit{ denarii} - 7\tfrac{1}{3}x = 52 \textit{ denarii}.$$

Therefore 44 *denarii*, which is the difference between 96 and 52, will equal $7\frac{1}{3}x$. So x equals 6 *denarii*, and this is the price of one foot of

[1] *Cf.* the discussion of this topic in Chapter LI of the *Practica Arithmeticae* (*Opera Omnia*, IV, 79–80): *Est etiam regula de Modo a me appellata, quoniam ex ipsa habentur regulae infinitae in rebus maxime mercantilibus, et potes replere librum ex ipsis in uno mense diversarum operationum, quae omnes regulae diversae videbuntur: et ita Frater Lucas, Borgias, Fortunatus, fecerunt libros per Neotericis instruendis, et ita tu lector poteris quotidie novas regulas et inusitatis fabricare.*

Modus est talis solve quaestionem quamvis per algebra deinde detrahe la co. et serva operationes easdem in terminis suis, et erit regula generalis.

green silk. Thus seven feet of the green cost 42 *denarii*, and three of the black amount to the difference between this and 72, namely 30 *denarii*, and one foot therefore is 10 *denarii*. Hence you have the price of both kinds of silk.

You have worked this by means of an unknown. Now I come to the rule and I say: Divide the greater length, namely 7 feet, and the number of *denarii*,[2] namely 72, by the smaller length, namely 3, and multiply the quotients by the number of feet assumed in the second case,[3] corresponding to the fewer, and from the product of the number of feet subtract the length remaining in the second case, and with the remainder divide the difference between the price, 2, and the product. The value of the greater length in the first case will result. For example, divide 7 and 72 by 3; the results are $2\frac{1}{3}$ and 24. Multiply [these] by 4, making $9\frac{1}{3}$ and 96. From $9\frac{1}{3}$ subtract 2, and from 96 subtract 52, leaving $7\frac{1}{3}$ and 44. Divide 44 by $7\frac{1}{3}$, and 6 results as the price of one foot of green silk.

From this there emerges this brief rule. As in the figure, divide 4 by 3 — i.e., by the number of feet of the same kind of silk in the two cases[4] — and $1\frac{1}{3}$ results. Multiply this by 7 and 72, making $9\frac{1}{3}$ and 96, from which subtract the figures given for the second case, which are 2 and 52, respectively, leaving $7\frac{1}{3}$ and 44. Divide the number of *denarii*, 44, by $7\frac{1}{3}$, the number of feet, and 6 is left as the price of one foot of green silk. In this way, from a long operation involving *x*, you can set up a very brief rule. Hence this rule of method may well be called the mother of rules.

Green	*Black*	*Price*
7 ft.	3 ft.	72 den.
2 ft.	4 ft.	52 den.

7		3		72
	$2\frac{1}{3}$		24	
		4		
	$9\frac{1}{3}$		96	
	2		52	
	$7\frac{1}{3}$		44	
		6		

2	4	52
7	3	72
$9\frac{1}{3}$	$1\frac{1}{3}$	96
$7\frac{1}{3}$	6	44

[2] 1545 has *numerum de*℞; 1570 has *numerum de* ℞; 1663 has *numerum rerum de* ℞.

[3] *in secunda positione.*

[4] The text has *in duabus petitionibus* = *in duabus positionibus.*

CHAPTER XXX

On the Golden Rule

This rule will cover a goodly share of those things which actually happen. First, having stated your problem in terms of an unknown and having perfected the operation, look for the closest solution [you can find]. Do it thus: Look for the whole numbers, greater and less, which most nearly satisfy the equation. These will not be difficult to discover. We will call the smaller of these the first approximation[1] and the greater the second approximation, and the difference between what they produce [i.e., the difference between the results when each of these is substituted on one side of the equation] we will call the great difference, the difference between that which the first produces and the constant of the equation we will call the first difference, and the difference between that which the second produces and the constant of the equation will be the second difference. Next divide the first difference by the great difference, and add the quotient to the first approximation. This gives us an imperfect solution which we substitute [for the unknown] in the equation[2] — that is, for the terms in the equation as [we did] with the first and second approximations. Subtract the results from the second product. Then subtract the imperfect solution from the second approximation and multiply the remainder by the second difference.[3] Divide this product by the difference between that which the second approximation produced and that which the imperfect solution produces. Subtract the quotient from the second approximation. The remainder is a close approximation to the value of x, to which by repeated operations it is always possible to approach still more closely. The same can be done where

[1] *primum inventum.* "First approximation" is hardly a happy translation of this phrase but it seems to be the best that can be found.

[2] *quam deducemus ad aequationem.*

[3] *residuum duc in differentiam secundum habitam.*

the equation consists of any power [equal] to a number and [other] powers, as will be seen in the examples.

Let, therefore, first

$$x^4 + 3x^3 = 100.$$

You see that if x is 2,

$$x^4 + 3x^3 = 40$$

and if x is 3,

$$x^4 + 3x^3 = 162.$$

Hence the first approximation is 2 and the first product 40; the second approximation is 3 and the second product 162; and 122 is the great difference, 60 the first difference, and 62 the second difference. (Note that the first approximation always differs from the second by unity, otherwise you have not performed the operation correctly.) Knowing all this, divide 60 by 122, and the result is $\frac{30}{61}$. Add this to 2, the first approximation, making $2\frac{30}{61}$ the imperfect solution. Raise this to the fourth power and to three times the cube, making about 85. Therefore, subtract 85, the product of the imperfect solution, from 162, the second product, and you will have 77. Subtract also $2\frac{30}{61}$ from 3, the second approximation, leaving $\frac{31}{61}$. Multiply by 62, the second difference, making $\frac{1922}{61}$.[4] Divide by 77 and $\frac{1922}{4697}$ results. Subtract this from 3, the second approximation. This, $2\frac{2775}{4697}$, will be a sufficiently close solution for

$$x^4 + 3x^3 = 100,$$

but if you wish, you can approximate x still more closely by further operations.

If

$$x^2 + 20 = 10x,$$

and if x equals 7, we will have

$$x^2 + 20 = 9\tfrac{6}{7}x,$$

and if x equals 8, we will have

$$x^2 + 20^{5} = 10\tfrac{1}{2}x.$$

Hence, as before, we have 7 as the first approximation and $9\frac{6}{7}$ as the first product, and we have 8 as the second approximation and $10\frac{1}{2}$ as

[4] 1570 and 1663 have $\frac{1922}{62}$.

[5] 1570 and 1663 have 27.

the second product, $\frac{9}{14}$ as the great difference, $\frac{1}{7}$ as the first difference, and $\frac{1}{2}$ as the second difference. We divide, therefore, the first difference by the great difference, and $\frac{2}{9}$ results. This we add to 7, the first approximation, making $7\frac{2}{9}$ the imperfect solution. The square of this plus 20 equals $9\frac{116}{117}x$. Therefore, since there is a nearly insensible difference between this and 10, the coefficient of x, we will not proceed any further but we will say that the solution is approximately $7\frac{2}{9}$.

Again, let us say that

$$x^3 = 6x + 20.$$

If x were 3, then $6x + 20$ would equal $1\frac{11}{27}x^3$ and if x were 4, $6x + 20$ would equal $\frac{11}{16}x^3$. Therefore, the first approximation is 3, the first product $1\frac{11}{27}$, the second approximation 4, and the second product $\frac{11}{16}$, the first difference $\frac{11}{27}$, the second difference $\frac{5}{16}$, the great difference $\frac{311}{432}$. With this last divide the smaller difference and $\frac{176}{311}$ results. Add this to 3, making the imperfect solution $3\frac{176}{311}$. Follow the equation through [with this value], viz., by applying[6] it to $6x + 20$, and these will be equal to $\frac{1,245,186,154}{1,363,938,029}x^3$. This, however, is approximately $\frac{31}{34}$.[7] From this we subtract the second product, and $\frac{61}{271}$[8] and $\frac{5}{16}$[9] will remain [the latter being the second difference]. Similarly, subtract $3\frac{176}{311}$, the imperfect solution, from 4, the second approximation, and $\frac{135}{311}$ remains. This I multiply by $\frac{5}{16}$, the second difference, as in the first example, making $\frac{675}{4976}$. Divide this by the difference between the second product and the product of the solution, which is $\frac{61}{271}$, and $\frac{182,925}{303,536}$ results. Subtract this from the second approximation as before, and the value of x is left as $3\frac{120,611}{303,536}$,[10] which is about $3\frac{201}{506}$, and therefore approximately $3\frac{2}{5}$, and

$$6x + 20 = 40\tfrac{2}{5}$$

while the cube of $3\frac{2}{5}$ equals $39\frac{38}{625}$. And, if you wish something still closer, you can go through the operation a third time in the same manner as you first did it and thus you will undoubtedly arrive at an insensible difference. This is universal reasoning and needs no other rule.

[6] 1545 and 1570 have *assumendo*; 1663 has *essumendo*.

[7] 1570 and 1663 have $\frac{31}{43}$.

[8] So in the text. It should, of course, be $\frac{61}{272}$. All subsequent calculations, however, are based on $\frac{61}{271}$.

[9] 1663 has $\frac{5}{26}$.

[10] 1570 and 1663 have $3\frac{120,611}{103,536}$.

And you will work similarly where there are three terms equal to two or three others. But with the arrival of two or three you can also reduce all to numbers, as in the first example, and the operations in this case are much easier. For example, if I say

$$x^4 + 6x^2 + 200 = 10x^3 + 12x,$$

the first approximation will be 9[11] and what it produces is -151,[12] this being the amount by which $10x^3 + 12x$ is greater than $x^4 + 6x^2 + 200$. And the second approximation will be 10 and what it produces is $+680$, since $x^4 + 6x^2 + 200$ are greater than $10x^3 + 12x$. Hence the first difference corresponds to[13] the first product and the second difference to the second product, and the great difference is the sum of them both. For the first operation, therefore, it is sufficient to divide, as before, the first difference by the great difference, and what results — it is $\frac{19}{104}$ — we add to the first approximation, thus making $9\frac{19}{104}$[14] the imperfect solution. Then if you desire to come still closer, you raise this solution to the proper powers on both sides and get the difference, which is called *A*. Multiply this by the difference between the imperfect solution and the second approximation and, again, divide the product by the great difference and add or subtract the quotient as is proper and you will have what was wanted. It is possible to work in the same way in the second and third examples, but we wished to show both methods so that whichever is easier in the circumstances [can be used]. The same can be said of the extraction of roots.

[11] 1570 and 1663 have 6.

[12] The text has -152.

[13] The text has *aequalis est.*

[14] The text has $9\frac{19}{304}$.

CHAPTER XXXI

On the Great Rule

This is a rule for solving great problems and from it were discovered the rules for alloys of gold and silver.[1] It incites ingenuity and does so through demonstration. It requires an expert and is [best] taught by problems since it is many-faceted. Commutation is the foundation of the rule.

PROBLEM I

Divide 8 into two parts the product of the cubes of which is 16. You say, therefore, that one part times the other is $\sqrt[3]{16}$. Divide 8 into two parts the product of which is $\sqrt[3]{16}$, and they will be

$$4 + \sqrt{16 - \sqrt[3]{16}}^{\,2} \quad \text{and} \quad 4 - \sqrt{16 - \sqrt[3]{16}}.$$

PROBLEM II

Divide 8 into three proportional parts the square of the first of which is equal to the others. First divide [8] into two parts the square of one of which is equal to the other. Then divide the larger [of these] into two parts in continued proportion with the less and they will be

$$\text{1st} \quad \sqrt{8\tfrac{1}{4}} - \tfrac{1}{2}$$

$$\text{2d} \quad \sqrt{\sqrt{631\tfrac{41}{64}} - 10\tfrac{3}{8}} + \tfrac{1}{4} - \sqrt{2\tfrac{1}{16}}$$

$$\text{3d} \quad 8\tfrac{1}{4} - \sqrt{2\tfrac{1}{16}} - \sqrt{\sqrt{631\tfrac{41}{64}} - 10\tfrac{3}{8}}^{\,3}$$

[1] *et ex ea inventae sunt regulae auri et argenti consolandi.* I read *consolidandi* for *consolandi.* The rules referred to are presumably those set out by Cardano in Chapter 41 of the *Practica Arithmeticae* (*Opera Omnia*, IV, 48 ff.), where he discusses problems of valuing alloys of gold and other metals.

[2] 1570 and 1663 have $4 + 16 + \sqrt[3]{16}$.

[3] The text has these values:

$$\text{1st} \quad \sqrt{8\tfrac{1}{4}} - \tfrac{1}{2}$$

PROBLEM III

Divide 8 into three proportional parts the square of the greatest of which is a proportional between the cubes of the others. You say, therefore, that the cube of the least is the cube root of the cube of the greatest and that, since the ratio of the cube of the greatest to its square is the greatest itself and this is also [the ratio] of the square of the greatest to the cube of the least, therefore, the cube of the least is the square root of the square of the greatest and is equal to the greatest. Hence 8 is the sum of the least and its cube and

$$x^3 + x = 8.$$

The value of x is the least.

PROBLEM IIII

Divide 8 into two parts such that seven times the greater is the proportional between the square of the greater and the cube of the less. Let A be the greater and C its square, and let B be the less and D its cube, and let E be $7A$. Since, therefore, A times A equals C, and A times 7 equals E, A will be to 7 as C is to E and hence, by V, 11 [of the *Elements*], as E is to D. Therefore AD equals $7E$. But E equals $7A$. Hence AD equals $49A$ and D equals 49, the square of 7. Hence the cube of B, the less, is 49 and B equals $\sqrt[3]{49}$ and A is the remainder.

PROBLEM V

Divide 8 into two parts such that seven times the greater is the proportional between the cube of the greater and the square of the less. Let A be the greater and let C equal A^3. Let B be the less and let D equal B^2. And let E equal $7A$. Since, therefore, A times A^2 equals C, and since A times 7 equals E,

$$A^2 : 7 = C : E$$

and, hence, as E[4] is to D. The ratio of A^2 to B^2, however, is composed of [i.e., is the product of] the ratio of A^2 to 7 and the ratio of 7 to B^2 and, therefore, of the ratio E to D and the ratio 7 to B^2. But D is B^2.

$$\text{2d} \quad \sqrt{\sqrt{631\tfrac{1}{4}} - 10\tfrac{3}{8}} + \tfrac{1}{4} - \sqrt{2\tfrac{1}{16}}$$

$$\text{3d} \quad 8\tfrac{3}{4} - \sqrt{18\tfrac{9}{16}} - \sqrt{\sqrt{631\tfrac{1}{4}} - 10\tfrac{3}{8}},$$

the internal radical sign being omitted in the last of these three terms.

[4] 1570 and 1663 omit E.

Hence the ratio A^2 to B^2 is composed of the ratios of $7A$ (which is E) to D and 7 to the same D. Therefore the ratio A^2 to D is composed of the ratios E to D and 7 to D. Hence, according to the rule for six quantities[5] or for the composition of proportions, $7E$ equals A^2D. But E is $7A$. Therefore $49A$ equals A^2D. Hence AD or AB^2 is 49. Hence, according to the rule for the cube and constant equal to the second power, B is $\sqrt{7\frac{1}{4}} + \frac{1}{2}$ and A[6] is $7\frac{1}{2} - \sqrt{7\frac{1}{4}}$.

Problem VI

Divide 8 into two parts such that their sum times the smaller is the proportional between [7]the product of the greater and the sum[7] and the product of the greater and the smaller. If, therefore, the smaller is multiplied by the greater and by the whole, the ratio of those products will be as the whole to the greater. Likewise, if the whole is multiplied by the greater and the smaller, [the ratio] of these products [will be] as the greater to the less. But the products are proportionals. Therefore, according to V, 11 of the *Elements*, the whole is to the greater as the greater is to the less. Hence 8 will be divided in accordance with a proportion having a mean and two extremes. Therefore the parts are manifestly $\sqrt{80} - 4$ and $12 - \sqrt{80}$.[8]

Problem VII

Divide 8 into two parts such that the product of the greater and the less is a proportional between the square of the less and ten times this same less. You say, therefore, since the less is that which is multiplied by itself, by the greater and by 10, that the greater is a proportional between the less and 10. Hence the square of the greater equals 10 times the less and the solution is known, for the greater will be $\sqrt{105} - 5$ and the less will be $13 - \sqrt{105}$.

Problem VIII

Divide 8 into two parts, the square of the greater of which is a proportional between the square of the smaller and the product of the whole and the greater. Let A be the greater and B the smaller. Since, then, $8A$ is a proportional between 64 and A^2, according to the demonstrations in our second [book] on Euclid, 64 will be a fourth proportional with the three products [i.e., with B^2, A^2, and $8A$].

[5] See Chapter 46 of the *Practica Arithmeticae* (*Opera Omnia*, IV, 67).

[6] 1570 and 1663 have 4 instead of A.

[7] 1663 omits these words.

[8] The text omits the last radical sign.

Hence

$$64:A^2 = A^4:B^4$$ [9]

and

$$8:A = A^2:B^2$$ [10]

according to VI, 17 of the *Elements*, for either is the mean of the ratios of its squares.[11] Hence A^3 equals $8B^2$, this having been demonstrated in the seventh book. Now let A [equal] the square [of C].[12] The cube of this [i.e., of C^2] will be the sixth power of the square root of A, [the latter] being C. Then B^2 equals one-eighth of C^6, and B equals the square root of one-eighth of C^6. Therefore, since the square root of a sixth power is a cube, B will equal $C^3\sqrt{\frac{1}{8}}$,[13] and since A is C^2,

$$C^2 + C^3\sqrt{\tfrac{1}{8}} = 8$$

and, having multiplied through by $\sqrt{8}$,

$$C^3 + C^2\sqrt{8} = \sqrt{512}.$$

Solve, therefore, in accordance with Chapter XV, working through the rules of the third book in the same manner as for rational numbers.[14]

Problem IX

Divide 8 into three proportional parts such that the sum of the first and second, the sum of the second and third, and 8 itself are proportionals. I say, first find the ratio among those proportional quantities the sum of the second and third of which is a proportional between the sum of the first and second and the sum of all three. Let these be the quantities A, B, and C. Since

$$(A + B + C):(B + C) = (B + C):(A + B)$$ [15]

from the statement of the problem, and

$$(B + C):(A + B) = C:B,$$

[9] *quare 64 ad quadratum A, ut quadrati A ad quadratum B duplicata.*

[10] *igitur 8 ad A, ut A ad B duplicata.*

[11] *nam utraque est media proportionum suorum quadratorum.*

[12] *quare ponemus A [aequalis C] quadratum.*

[13] *erit B aequalis cubi C parti* ℞ $\frac{1}{8}$.

[14] 1545 has *ut in numeris rationalibus operando*; 1570 and 1663 have *ut in numeris notis ac veris operando.*

[15] The text omits all indications of addition here and elsewhere in this problem.

according to V, 12 of the *Elements*,

$$(A + B + C):(B + C) = C:B$$[16]

according to the 11th proposition of the same. But B times the ratio is C and, therefore, $B + C$ times the ratio is $A + B + C$. Now let[17] D be the product of the ratio and C. Hence, since the ratio times B is C and the same times C is D, the ratio times $B + C$ will be $C + D$ and the same times $B + C$ will be $A + B + C$. Therefore $A + B + C$ equals $C + D$. Subtract C, and $A + B$ is left equal to D. D, however, is a fourth proportional quantity. Therefore four quantities in continued proportion are to be found, the fourth of which is equal to the sum of the first two. Assume, then, that the first is 1, the second x, the third x^2, and the fourth x^3. Hence

$$x^3 = x + 1,$$

and the value of x, which is a proportional, is known from this equation. Divide, therefore, 8 into four quantities in accordance with this continued proportion, as is shown in the sixth book. We have also solved this problem in another manner in the fourth book.

Problem X

Divide 8 into two parts, seven times the greater of which is a proportional between the cube of the smaller and the product of the greater and smaller. Let A be the smaller, C its cube, B the greater, E the product of B and A, and D seven times B. [18]Since, therefore, BA is E and $7B$ is D,[18]

$$A:7 = E:D,$$

whence

$$A:7 = D:C.$$

Therefore AC equals $7D$. But D is $7B$. Hence $49B$ equals A^4 and, therefore, B equals $\frac{1}{49}A^4$. Since, then,

$$A + B = 8$$

and B equals $\frac{1}{49}A^4$,

$$A + \tfrac{1}{49}A^4 = 8,$$

[16] The text has $B:C$.

[17] Reading the *ut* which appears in the text as a misprint for *sit*.

[18] 1570 and 1663 omit this much of the sentence.

wherefore

$$x + \frac{1}{49}x^4 = 8.^{19}$$

Therefore[20]

$$x^{4\ 20} + 49x = 392.$$

Now, although the rule for this is not general, yet it leads to a very pretty [solution] of this problem.

2. This [i.e., the "great rule"] should be tried whenever it is impossible to solve a given problem by any of the usual methods, as is easily seen.[21]

[19] 1663 omits "= 8." *Quaere* why Cardano suddenly switches back to x at this point.

[20] 1663 omits these words.

[21] 1545 has *Deprehenditur et quandoque impossibilitas eodem modo propositarum quaestionum, ut facile est videre*; 1570 and 1663 have *Deprehenditur et quandoque eodem modo quod propositae quaestiones sint impossibiles.*

CHAPTER XXXII

On the Rule of Equal Position[1]

1. This rule is more useful than [that of] simple position in all problems in which there are parts to be multiplied in the same manner, poorer where they are to be multiplied dissimilarly.[2] In the latter, simple position is the easier. Thus if I should say, Divide 8 into two parts one of which multiplied by the square or cube of the other gives 20, you would arrive at

$$8x^2 - x^3 = 20$$

by simple position or, in the second case, at

$$8x^3 - x^4 = 20.$$

By assuming [the two parts to be] $4 + x$ and $4 - x$, you would arrive at

$$16x + 44 = x^3 + 4x^2$$

or, in the second case, at

$$128x +^{3} 236 = x^4 + 8x^3.$$

It is evident, therefore, that the latter [equations] are more difficult than the former.

Furthermore, [when we use] simple position we discover the value of x in the first operation, [but] in the case of equal position [we discover]

[1] *De regula aequalis positionis. Positio* here and in some of the following chapters is a bothersome word. For lack of anything better and after a good deal of hesitation, I have borrowed the word "position" from the phrase "the rule of false position" (see David Eugene Smith, *History of Mathematics*, II (Boston, 1925), 437 ff.) as a translation for it, even though "position" is no longer a familiar one in the sense in which it is used there and here. Other English words, however — for example, "assumption," "supposition," "positing," etc.—do not carry with them quite the right flavor. If the reader finds it easier for himself to substitute one of these for "position," he can be assured that he will not be wrong.

[2] *Haec regula est utilior positione simplici, in omnibus quaestionibus ubi partes aequaliter multiplicantur, secus ubi inaequaliter.*

[3] 1570 and 1663 have *7* instead of *et* or *&*.

a difference which is to be added to or subtracted from one-half the dividend to obtain the sought-for numbers which are the value of x, although we [also] assumed the unknown [itself] to be a difference.

I call it [a case of] simple position when I say, Divide 10 into two parts the product of which is 20 and assume one part to be x, the other to be $10 - x$. But [I call the case one of] equal position if I say that one is $5 + x$ and the other $5 - x$. Hence, since the simple case is already familiar, the case of equal position will be discussed through problems and examples, since its most frequent use and utility [lies here].

Problem I

There is a triangle the difference between the first and second sides of which is 1 and between the second and third sides of which is also 1, and the area of which is 3. Now assume x to be the second side. The first side will be $x - 1$ and the third $x + 1$. Follow the rule for triangles given in the following book, and this makes $\sqrt{\frac{3}{16}x^4 - \frac{3}{4}x^2}$ equal to 3. Therefore

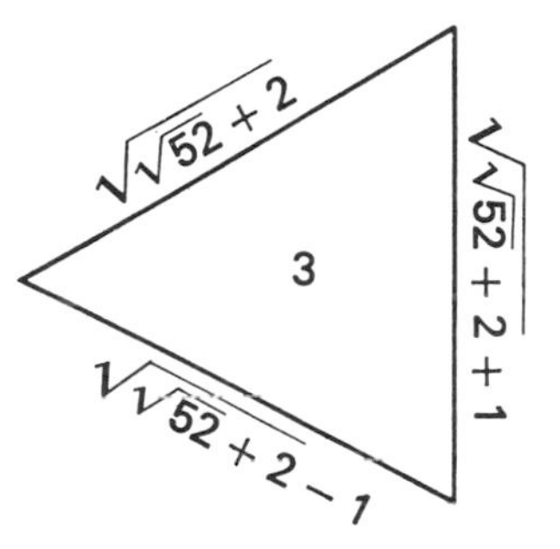

$$\tfrac{3}{16}x^4 = \tfrac{3}{4}x^2 + 9$$

and, therefore,

$$x^4 = 4x^2 + 48,$$

and x will be, from the rule for derivatives, $\sqrt{\sqrt{52} + 2}$, and this is the second side. Add and subtract 1, therefore, and you will have the remaining sides, as shown in the figure.

Problem II

Divide 10 into two parts the sum of the cubes and squares of which is 400. Let the first part be $5 + x$ and the second $5 - x$. Follow through with the problem by raising the parts to the cube and the square. Then, some of the parts having dropped out, collect [the rest], making

$$32x^2 + 300 = 400.$$

Hence[4] x^2 equals $3\frac{1}{8}$ and the unknown, which is the difference, equals $\sqrt{3\frac{1}{8}}$. Therefore the parts are $5 + \sqrt{3\frac{1}{8}}$ and $5 - \sqrt{3\frac{1}{8}}$.

[4] 1545 and 1663 have *quare*; 1570 has *quere*.

Problem III

Divide 6 into two parts the sum of the squares of which is equal to the difference between their cubes. Let the greater by $3 + x$ and the smaller $3 - x$. Follow the problem through and you will have as the sum of the squares $2x^2 + 18$ and as the difference between the cubes $2x^3 + 54x$, and these are equal to each other. Hence

$$x^3 + 27x = x^2 + 9.$$

Follow the rule and the value of the unknown — i.e., of the difference — appears as $\sqrt[3]{\sqrt{702\frac{1}{3}} + \frac{1}{27}} - \sqrt[3]{\sqrt{702\frac{1}{3}} - \frac{1}{27}} + \frac{1}{3}$. Therefore the parts will be $3\frac{1}{3} + \sqrt[3]{\sqrt{702\frac{1}{3}} + \frac{1}{27}} - \sqrt[3]{\sqrt{702\frac{1}{3}} - \frac{1}{27}}$ [as the greater] and, as the less, $2\frac{2}{3} + \sqrt[3]{\sqrt{702\frac{1}{3}} - \frac{1}{27}} - \sqrt[3]{\sqrt{702\frac{1}{3}} + \frac{1}{27}}$.

Problem IIII

Divide 8 into two parts the product of the greater and smaller of which is a proportional between the smaller and nine times the greater. Let the first be $4 + x$ and the smaller $4 - x$. Follow through with what was proposed and nine times the greater will be $36 + 9x$, the greater times the smaller will be $16 - x^2$, and the smaller will be $4 - x$. These are proportionals.[5] The product of $36 + 9x$ and $4 - x$ equals $(16 - x^2)^2$. In multiplying $36 + 9x$ by $4 - x$, the first power drops out. Hence the product is $144 - 9x^2$, which is equal to $(16 - x^2)^2$ which is $256 + x^4 - 32x^2$. Therefore, by setting the negatives over to the other side,

$$112 + x^4 = 23x^2$$

and you will have x equal to $\sqrt{11\frac{1}{2} - \sqrt{20\frac{1}{4}}}$, which is $\sqrt{7}$. Add this to or subtract it from 4, and the parts sought for will be $4 + \sqrt{7}$ and $4 - \sqrt{7}$.

If you will, you can solve this by simple [position] arriving at an equation of the cube and first power equal to the square and constant. Yet the operation is still quite intricate [and] beyond comparison [with that under the other method], for you need more than ten additional operations before you can arrive at the true solution which is always in the nature of a *binomium* or a true *recisum*, not an improper one.

[5] 1545 has *et haec sunt proportionalia*; 1570 and 1663 have *et haec sunt in eadem proportione.*

Problem V

Divide 10 into two parts such that the sum of the square roots of 100 minus the square of the first and 97 minus the square of the second is 17. If you wish to avoid work, you will first explore tentatively whether the case is possible. Knowing that it is, then, let the first part be $5 + x$ and the other $5 - x$. Square each of these and subtract the square of the greater from 100 and that of the less from 97, and you will have as remainders $75 - x^2 - 10x$ and $72 - x^2 + 10x$, the sum of the square roots of which must equal 17. Hence 17 minus one of these roots equals the other. So we square $17 - \sqrt{75 - x^2 - 10x}$ and have $364 - x^2 - 10x - \sqrt{86{,}700 - 1156x^2 - 11{,}560x}$ equal to the square of the other root, namely $72 - x^2 + 10x$. Delete the likes from both sides and take the single universal root to the other side, as we have shown in the fourth and fifth books,[6] and you will have

$$292 - 20x = \sqrt{86{,}700 - 1156x^2 - 11{,}560x}.$$

Hence, by squaring the parts again, you will have

$$86{,}700 - 1156x^2 - 11{,}560x = 85{,}264 + 400x^2 - 11{,}680x.$$

Pursue this to a solution by reducing to a single x^2 and you will have $\sqrt{\frac{139{,}876}{151{,}321}} + \frac{15}{389}$ as the value of x. But $\sqrt{\frac{139{,}876}{151{,}321}}$ is $\frac{374}{389}$. Therefore, by adding $\frac{15}{389}$ to it, it becomes $\frac{389}{389}$. Hence x equals 1, and the parts are 4 and 6.

Problem VI

Equal position cannot solve every problem and simple position can. For example: Make two parts of 8 the square of the greater of which is a proportional between the product of the greater and the smaller and $7\frac{1}{2}$ times the whole,[7] namely 60. Let x be the greater and you will have 60, x^2, and $8x - x^2$ as the proportionals. Wherefore, by multiplying the mean by itself, you will have

$$x^4 = 480x - 60x^2.$$

Depress these and you will have

$$x^3 + 60x = 480$$

and therefore x is known. Through equal position, however, you would arrive at an equation consisting of five terms which could be solved through the great rule. But this does not pertain to the present subject.

[6] 1570 and 1663 have third and fourth books.

[7] The text has *decuplum totius* (10 times the whole). How did *sesquiquintuplum* get turned into *decuplum*?

CHAPTER XXXIII

On the Rule of Unequal or Proportional Position[1]

This rule teaches us how, given unequal numbers, we may annex unequal quantities to both alike[2] in such fashion that, through cross-multiplication, similar parts will cancel out. I will show this by examples, although problems which are solved by this method can also be solved by the rule for working backward to the unknown which was discussed in Chapter V.

PROBLEM I

Example: There are two numbers, the difference between which is 4 and the square of the smaller of which plus the square of one-half the greater plus the square root of the sum of these squares is 110. By working backward you could say, therefore, that 110 is composed of the sum of the squares and the square root of the sum. Hence, letting x^2 be the sum,

$$110 = x^2 + x,$$

wherefore x equals 10 and x^2 equals 100. Therefore you divide 100 into two parts twice the square root of one of which is greater than the square root of the other by 4. The solution is clear.

We proceed thus under the present method: Let the first [or] smaller number be $2x$ — [I use $2x$ rather than x] since the part [of the greater we deal with] is one-half — and the greater will be $2x + 4$. Then take that part, or one-half, of the second which is to be squared. This will be $x + 2$, and the first number, as was said, is $2x$. Having

[1] *De regula inaequaliter ponendi, seu proportionis.* In the 31st and 32d problems in the *Ars Magna Arithmeticae*, this "rule" is also referred to as the *regula positionis proportionatae* and as *positio proportionalis*.

[2] *positiones pariter aequales annectamus.* The context and examples make it clear that *inaequales* is meant rather than *aequales*.

gone thus far,[3] it is possible,[4] without changing the nature of the problem, to combine each part of the number with the x terms in such fashion that, when squared, the first powers drop out. You proceed in this way: Consider what part of the first number the second — i.e., the number that is to be squared — is. In the example, what part of $2x$ is $x + 2$? You discover that it is $\frac{1}{2}$ [with a remainder of] 2.[5] Now square the denominator and numerator of the fraction [i.e., $\frac{1}{2}$] and add the results and you will have 5 for a divisor. Then square the numerator and multiply the result by the difference between the numbers, which is 4, making 4 again, for a dividend. Divide 4 by 5, and $\frac{4}{5}$ results. Subtract this from $2x$, or the greater [sic] part, and you will have $2x - \frac{4}{5}$. Then divide $\frac{4}{5}$ by $\frac{1}{2}$,[6] and $\frac{8}{5}$ results. Add this to x and you will have $x + \frac{8}{5}$. Lo, you now see that, since you have $2x - \frac{4}{5}$ and $x + \frac{8}{5}$, the ratio of $\frac{8}{5}$ to $\frac{4}{5}$ is as $2x$ to x and, if you will take twice the greater, namely $2x + 3\frac{1}{5}$, it will exceed the smaller, namely $2x - \frac{4}{5}$, by exactly 4. This being completed, square these parts in accordance with the general rule, and you will have $4x^2 + \frac{16}{25} - \frac{16}{5}x$ and $x^2 + 2\frac{14}{25} + \frac{16}{5}x$. Add these and you will have $5x^2 + 3\frac{1}{5}$. This plus its square root equals 110. Therefore the root is equal to 110 minus the sum, or

$$106\tfrac{4}{5} - 5x^2 = \sqrt{5x^2 + 3\tfrac{1}{5}}.$$

Square the parts and you will have

$$5x^2 + 3\tfrac{1}{5} = 11{,}406\tfrac{6}{25} + 25x^4 - 1068x^2.$$

Set the negative on the other side and divide by the coefficient of x^4, which is 25, and you will have

$$x^4 + 456\tfrac{76}{625} = 42\tfrac{23}{25}x^2,$$

wherefore x equals $\sqrt{21\frac{23}{50} - \sqrt{4\frac{1025}{2500}}}$. But $\sqrt{4\frac{1025}{2500}}$ is $2\frac{1}{10}$. Therefore x is $\sqrt{19\frac{36}{100}}$, which is $4\frac{2}{5}$. Hence x equals $4\frac{2}{5}$. But the first or greater part was $2x - \frac{4}{5}$,[7] hence it is 8. The smaller was $x + \frac{8}{5}$ and is, therefore, 6, and twice this is 12, which is greater than 8 by 4, which is what we wanted.

[3] *hoc habito.*

[4] Reading *possibilis* for the *positum* of the text.

[5] *invenies quod est* $\frac{1}{2}$ *p: 2.*

[6] *Deinde divide* $\frac{4}{5}$ *per* $\frac{1}{2}$ *partem.*

[7] 1570 and 1663 have $\frac{5}{4}$.

Problem II

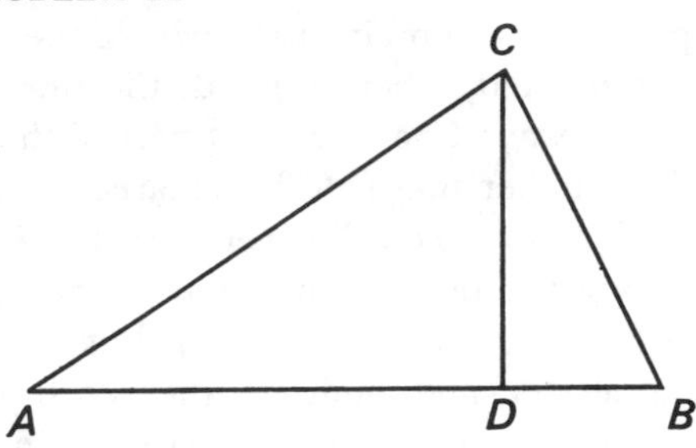

There is a triangle, ABC, the base of which, AB, is 8 plus its perpendicular CD, and AD is 3 DB, and BC^2 plus the side CB is 182. Let CD be x and AB $x + 8$; or let CD be $4x$ and AB $4x + 8$. BD will then be $x + 2$, and the ratio [between it and AB] will be $\frac{1}{4}$. Hence, as before, square 4, making 16. Square 1, making 1. Add them, making 17 the divisor. Then square 1, the numerator of $\frac{1}{4}$, and 1 results. Multiply this by 8, the difference, and 8 results. Divide by 17, making $\frac{8}{17}$ as the part to be subtracted from $4x$. Then divide $\frac{8}{17}$ by the ratio, which is $\frac{1}{4}$, and $\frac{32}{17}$ results as the part to be added to x. BD, therefore, will be $x + \frac{32}{17}$ and CD will be $4x - \frac{8}{17}$. Square these parts and you will have the squares of CD and BD equally derived. Consequently BC^2 is $17x^2 + 3\frac{221}{289}$. Follow through as in the preceding by adding to it the side BC and

$$\sqrt{17x^2 + 3\tfrac{221}{289}} + 17x^2 + 3\tfrac{221}{289} = 182.$$

Therefore

$$178\tfrac{68}{289} - 17x^2 = \sqrt{17x^2 + 3\tfrac{221}{289}}.$$ [8]

Perform this operation, therefore, as before, and you will have $3\frac{2}{17}$ as the value of x. Since, then, BD equals $x + \frac{32}{17}$, BD will equal 5, and AB will equal 20 or $4BD$. Hence CD, which is 8 less than AB, will be 12.

Problem III

Similarly, if it should be said that there are two numbers the difference between which is 12 and that the square of the less plus the square of three-tenths the greater plus the square of this sum equals 1000, you would proceed as before by squaring the numerator and the denominator [of $\frac{3}{10}$] and adding them. Thus 109 is the divisor. Then I square 3, the numerator, and multiply the product by 12, making 108. I divide this by 109, and I have the part to be subtracted from $10x$. Then I divide $\frac{108}{109}$[9] by $\frac{3}{10}$, making $\frac{360}{109}$, the part to be added to $3x$.

[8] 1663 has $7x^2$.
[9] 1570 has $\frac{108}{209}$.

If, therefore, $3x + \frac{360}{109}$[10] is multiplied by $\frac{10}{3}$, the product will be 12 more than $10x - \frac{108}{109}$, and the ratio of $\frac{360}{109}$ to $\frac{108}{109}$ is the same as 10 to 3.

Now these rules have infinite forms. Thus, if we should say that the difference between one-half of one number and one-third of another is 12, and the sum of their squares and their square roots is 100, then look for the proper rule in the same way [or] by the rule of method since the present rule is a branch of that one. Look first for two numbers which differ by 12, one-half the first of which is so divided into $\frac{1}{2}x$ plus a number and [one-third] the other into $\frac{1}{3}x$ plus a number that the products of the unknown are equal. Assume one number to be positive and the other negative and proceed according to the ninth chapter. For a surd, as in these, I give another rule in the [next] example: I say there whatever can be said [about it].

PROBLEM IIII

Find two numbers the difference between which is 14 and the square of one-third of one of which plus the square of one-fourth the other plus the square root of the sum of these squares is 110. I say: First multiply the denominators by the numerators in turn[11] — i.e., 4 by 1, and 3 by 1 — and add the squares of their products, these being 4 and 3, and you will have 25 as a divisor. Then multiply the denominators, 3 by 4, making 12, and multiply this by 14, the [given] difference, and 168 results. Multiply this by the product of the numerators, which is 1, and 168 again results. For the dividend, therefore, divide 168 by 25, making $\frac{168}{25}$. Multiply this by the given parts, $\frac{1}{3}$ and $\frac{1}{4}$, and you will have $2\frac{6}{25}$ as that which is to be added and $1\frac{17}{25}$ as that which is to be subtracted since, as has always been said, the smaller part of a number is subtracted from the greater and the greater is added to the smaller. Multiply, then, $\frac{1}{3}x - 1\frac{17}{25}$ by itself and likewise $\frac{1}{4}x + 2\frac{6}{25}$, and collect the products. You will have $\frac{25}{144}x^2$ and $7\frac{103}{125}$ without any first power, Hence you follow the process as in the preceding cases.

Another example of little difficulty in this rule: Find two numbers differing by 4, the square of three-fourths the smaller of which plus the square of two-thirds the greater plus the sum of the products and their roots makes 110. Therefore you cross-multiply[12] 3 and 3, and 4 and 2,

[10] 1570 has $\frac{360}{209}$.

[11] 1545 and 1570 have *duc denominatores in numeratores*; 1663 has *duc nominatores in nominatores*.

[12] *duces igitur in crucem.*

making 9 and 8, the sum of the squares of which is 145, the divisor. Likewise, you multiply the denominators, 3 and 4, making 12. Multiply this by 4, the difference in the numbers, making 48, and multiply this by 6, the product of the numerators, making 288 the dividend. Then having divided 288 by 145, $\frac{288}{145}$ results. Multiply this by $\frac{2}{3}$ and $\frac{3}{4}$ separately, the parts to be taken, and you will have $\frac{192}{145}$ and $\frac{216}{145}$, the parts to be added or subtracted as before.

PROBLEM V

We may say the same of a sum as, for example, if it should be said, Divide 10 into two parts the square of one of which plus the square of half the other plus the square root of the whole is 30. I say that you will proceed according to the rule given in the first problem, since this is [made up] of integers on the one side. You therefore discover the numbers 4 and 2, and from the greater you subtract x and to the smaller you add $2x$. In this [respect the present rule] differs from the rules for the difference between numbers; in other [respects] they are the same. Therefore, by continuing the operation, you will have $\sqrt{2\frac{1}{10} - \sqrt{\frac{121}{100}}}$, which is to say 1, as the value of x. Hence the numbers are 6 and 4.

CHAPTER XXXIIII

On the Rule for a Mean[1]

1. This rule is thus named by me because a mean or ratio [between two numbers] is sought and because, in order to avoid the confusion of unity, we assume one part to be half of unity. Its only use is in finding quantities which are multiplied equally and which serve [as members of] a proportion. When they do not serve in this way, the use of this rule is not helpful. It will be explained only in [terms of] two quantities; we will speak of more [than two quantities] in Chapter XXXIX. It is evident, moreover, that if someone says, Find two numbers the sum of the squares of which [2]is 10 and the sum of the cubes of which[2] is 30, this rule will not be of service, since the ratio of 30 to 10, which is triple, cannot hold between cubes and squares, their quantity being different. But it will be useful to show by examples how this rule and others [like it work].

Problem I

Find two numbers such that the product of their differences and the difference between their squares is 10[3] and such that the product of their sum and the sum of their squares is 20. Let one of these be x, as

[1] In the *Practica Arithmeticae* (*Opera Omnia*, IV, 87), Cardano gives us this information about the origin of the Rule for a Mean: *Et hanc habui a Magistro Gabriele de Aratoribus, qui eam habuit a Fratre Luca, et est ingeniosa valde his autem Magister Gabriel fuit ille qui impulit me ut componerem hunc librum . . . regula autem dupli a me inventa est facilior quidem sed plura praesupponit. Haec autem si per se sumatur longe est universalior.* The same Gabriel appears elsewhere in the same work (*Opera Omnia*, IV, 78) in a discussion of dividing 10 by $3 - \sqrt[3]{5}$: *et hanc regulam habui a Magistro Gabriele de Aratoribus Arithmeticam Mediolani publice docente. Cf.* Moritz Cantor, *Vorlesungen über Geschichte der Mathematik* (2d ed., Leipzig, 1900), II, 482.

[2] 1663 omits these words.

[3] 1663 has 20.

has been said, and let the other be $\frac{1}{2}$.[4] Then find their difference and their sum and the squares of the parts and the difference and sum of the squares, as in the margin. Multiply the difference between the parts by the difference between the squares and you will have $\frac{1}{8} + x^3 - \frac{1}{2}x^2 - \frac{1}{4}x$, and this must be one-half the product of the sums of the numbers and their squares, since 10 is half of 20. That is, it will be one-half of $\frac{1}{8} + x^3 + \frac{1}{2}x^2 + \frac{1}{4}x$. Therefore

$$\frac{1}{4} + 2x^3 - x^2 - \frac{1}{2}x = \frac{1}{8} + x^3 + \frac{1}{2}x^2 + \frac{1}{4}x.$$

Numbers	x	$\frac{1}{2}$
Difference between the numbers	$x - \frac{1}{2}$	
Sum of the numbers	$x + \frac{1}{2}$	
Squares [of the numbers]	x^2	$\frac{1}{4}$
Difference between the squares	$x^2 - \frac{1}{4}$	
Sum of the squares	$x^2 + \frac{1}{4}$	
Product of the differences	$\frac{1}{8} + x^3 - \frac{1}{2}x^2 - \frac{1}{4}x$	
Product of the sums[1]	$\frac{1}{8} + x^3 + \frac{1}{2}x^2 + \frac{1}{4}x$	

$$\frac{1}{4} + 2x^3 - x^2 - \frac{1}{2}x$$
$$\frac{1}{8} + x^3 + \frac{1}{2}x^2 + \frac{1}{4}x$$
$$\frac{1}{8} + x^3 = 1\frac{1}{2}x^2 + \frac{3}{4}x$$

$x + \frac{1}{2}$	$x + \frac{1}{2}$
$x^2 - \frac{1}{2}x + \frac{1}{4}$	$1\frac{1}{2}x$

[1] The text has "product of the squares."

Then, by converting the parts from minus to plus, we will have

$$\frac{1}{8} + x^3 = 1\frac{1}{2}x^2 + \frac{3}{4}x.$$

Wherefore, having divided the parts in order to facilitate the operation, which should always [be done] when they can be divided, we will have

$$1\frac{1}{2}x = x^2 - \frac{1}{2}x + \frac{1}{4}$$

since the divisor is made up of the parts assumed at the beginning, namely $x + \frac{1}{2}$. Therefore

$$x^2 + \frac{1}{4} = 2x$$

and x equals $1 + \sqrt{\frac{3}{4}}$. The two quantities, then, are in the ratio of $1 + \sqrt{\frac{3}{4}}$ to $\frac{1}{2}$ and therefore in the ratio of $2 + \sqrt{3}$ to 1.

Now we look for two quantities in this ratio the sum of which multiplied by the sum of their squares is 20, for such is necessary to meet the remaining condition [of the problem]. Let one of them be y and the other $y(2 + \sqrt{3})$, and we find their squares, which we add.

[4] 1570 has $\frac{3}{2}$.

The sum is $(8 + \sqrt{48})y^2$. We multiply this by the sum of the numbers, namely $(3 + \sqrt{3})y$, making $(36 + \sqrt{1200})y^3$. We divide, therefore, 20 by $\sqrt{1200} + 36$, and the result is $7\frac{1}{2} - \sqrt{52\frac{1}{12}}$, the cube root of which is the smaller number that was to be found. The greater will be had by multiplying the smaller by $2 + \sqrt{3}$. Therefore the numbers which were to be found will be

$$\text{1st: } \sqrt[3]{7\tfrac{1}{2} - \sqrt{52\tfrac{1}{12}}}$$

$$\text{2d: } \sqrt[3]{195 - \sqrt{35{,}437\tfrac{1}{2}} + \sqrt{37{,}968\tfrac{1}{4}}\text{[5]} - \sqrt{35{,}490}}.$$

Problem II

Find two numbers such that their difference times the difference between their cubes is 10 and their sum times the sum of their cubes is 30. In this problem you proceed as in the preceding but, to make it easier, assume the parts to be x and 1. Follow through as in the preceding problem until you come to

$$x^4 + 1 = 2x^3 + 2x.$$

There are now five quantities in continued proportion, the sum of the first and fifth of which is twice the sum of the second and fourth. Therefore I seek the ratio in accordance with the rule for five quantities in continued proportion, by assuming 2 and 4, for instance, of which 4 is twice the other, and by dividing 4 between the first and fifth and 2 between the second and fourth. Therefore the ratio will be that of $\frac{1}{2} + \sqrt{\frac{3}{4}} + \sqrt{\sqrt{6\frac{3}{4}} - 2\frac{1}{4}} + \sqrt{\sqrt{\frac{3}{4}}\text{[6]} - \frac{3}{4}}$ to 1. Assume, therefore, y with these as coefficients, namely y and

$$y\left(\tfrac{1}{2}\text{[7]} + \sqrt{\tfrac{3}{4}} + \sqrt{\sqrt{6\tfrac{3}{4}} - 2\tfrac{1}{4}} + \sqrt{\sqrt{\tfrac{3}{4}}\text{[8]} - \tfrac{3}{4}}\right).$$

Then cube these parts following the rules in the third book. This is not difficult. Then multiply $y\left(\sqrt{\frac{3}{4}} + \sqrt{\sqrt{6\frac{3}{4}} - 2\frac{1}{4}\text{[9]}} + \sqrt{\sqrt{\frac{3}{4}}\,[-\frac{3}{4}]} - \frac{1}{2}\right)$, the difference between the numbers, by the difference between their cubes, which can be had by subtracting x^3 from the cube of the aforesaid four terms, and the product will equal 10. Divide 10 by this product and the fourth root of the quotient will be the value of the

[5] The text has $\sqrt{33{,}075}$.

[6] 1570 and 1663 have $\frac{1}{4}$.

[7] 1570 and 1663 have $\frac{3}{4}$.

[8] 1570 and 1663 omit the internal radical sign.

[9] 1663 has $2\frac{1}{2}$.

first quantity. This multiplied by $\frac{1}{2} + \sqrt{\frac{3}{4}} + \sqrt{\sqrt{6\frac{3}{4}} - 2\frac{1}{4}} + \sqrt{\sqrt{\frac{3}{4}} - \frac{3}{4}}$ yields the second quantity or number.

Problem III

Find two numbers the sum of the fifth powers of which is 20 and the sum of the cubes of which times the sum of the squares is 25. Assume the parts to be x and 1, as in the preceding, and their fifth powers are x^5 and 1, and the product of the sum of their squares and cubes is $x^5 + x^3 + x^2 + 1$. This is to $x^5 + 1$ as 25 is to 20 and as 5 is to 4. Therefore by the rule for proportional quantities multiply x^5 [10] $+ x^3 + x^2 + 1$ by 4. We get the same by multiplying $x^5 + 1$ by 5. Hence

$$4x^5 + 4x^3 + 4x^2 + 4 = 5x^5 + 5.$$

Hence, having subtracted, we will have

$$x^5 + 1 = 4x^3 + 4x^2.$$

Divide the parts by $x + 1$, and

$$x^4 - x^3 + x^2 - x + 1 = 4x^2.$$

Therefore

$$x^4 + 1 = x^3 + 3x^2 + x.$$

Hence there are five quantities in continued proportion, the sum of the first and fifth of which is, by way of example, 10, and the sum of the second and fourth plus three times the third is also 10. Therefore, the ratio will be known by the rule for five quantities in continued proportion, and it will be $\dfrac{\sqrt{3\frac{6}{7}} - 1 + \sqrt{\sqrt{1\frac{5}{7}} - \frac{6}{7}}}{2 - \sqrt{1\frac{5}{7}}}$ and this is the ratio of these quantities. In the second stage,[11] therefore, assume x, and x multiplied by the aforementioned number or ratio, as in the preceding example. Having divided by the numerator, raise the reduced proportion to the fifth power, by the proper rule; add to this 1, the fifth power of 1; and divide 20 by this sum. The fifth root of the quotient is the smaller number. Multiply this by the ratio and the greater will appear. To put through such an operation is something almost beyond human ability and would be entirely impossible except for the rules of the third book.

[10] 1570 and 1663 have $2x^5$.
[11] *in secunda . . . positione.*

CHAPTER XXXV

On the Rule for a Sum

1.[1] In this case it is a sum that we would have just as, under the preceding rule and the iterative rule, it is a ratio that is sought.[2] [The present rule,] moreover, is extremely useful when no ratio between parts is given. For the means of discovering several numbers simultaneously is either a ratio or a sum or a difference. Whereas, therefore, in the preceding and iterative rules a ratio is to be had, in this one, on the other hand, a sum or difference is enough. Hence this antedated the others by some time. We can also call this the rule of the double, because it involves two parts or numbers which are being sought. Its rationale, like that of the others, is shown by examples.

Problem I

Find two numbers the sum of the squares of which is 20 and the product of the two of which plus the numbers themselves is 10. I say that, although this can be solved in accordance with the sixth book, we will work it out according to this rule. Let the sum be x or the unknown. Since the product of the two is 10 minus their sum, it will be $10 - x$. Divide x, therefore, into two parts the product of which is $10 - x$, and these will be, according to the rule given in the chapter on operations in the sixth book, $\frac{1}{2}x + \sqrt{\frac{1}{4}x^2 + x - 10}$ and $\frac{1}{2}x - \sqrt{\frac{1}{4}x^2 + x - 10}$. The sum of the squares of these ought to be 20 and, since one part is a *binomium* and the other a *recisum* with respect to $\frac{1}{2}x$, it suffices to multiply each part [of the *binomium* and *recisum*] by itself, not one by the other, as we have shown in books three, four,

$$\begin{array}{c} \frac{1}{2}x + \sqrt{\frac{1}{4}x^2 + x - 10} \\ \frac{1}{2}x + \sqrt{\frac{1}{4}x^2 + x - 10} \\ \hline \frac{1}{4}x^2 + \quad \frac{1}{4}x^2 + x - 10 \\ \hline \frac{1}{2}x - \sqrt{\frac{1}{4}x^2 + x - 10} \\ \frac{1}{2}x - \sqrt{\frac{1}{4}x^2 + x - 10} \\ \hline \frac{1}{4}x^2 + \quad \frac{1}{4}x^2 + x - 10 \\ \hline x^2 + 2x - 20 \end{array}$$

[1] Omitted in 1663.

[2] *Sicut ex praecedente, et regula iterata, proportio ipsa quaeritur, sic per hanc habemus aggregatum.* The "iterative rule" or "iterative method" is illustrated in Chapter XLIX

and five. The square of $\frac{1}{2}x$ is $\frac{1}{4}x^2$, and the square of $\sqrt{\frac{1}{4}x^2 + x - 10}$ is $\frac{1}{4}x^2 + x - 10$, and the other part [produces] the same, as [shown] in the figure. Hence the squares of the *binomium* and *recisum* are $x^2 + 2x - 20$, and this is equal to 20, as has been said. Therefore

$$x^2 + 2x = 40,$$

wherefore x will equal $\sqrt{41} - 1$. Divide $\sqrt{41} - 1$ into two parts, the sum of the squares of which is 20. There will be [derived] through a fresh unknown or according to the rules in the chapter on operations in the sixth[3] book these parts: (1) $\sqrt{10\frac{1}{4}} - \frac{1}{2} + \sqrt{\sqrt{10\frac{1}{4}} - \frac{1}{2}}$; and (2) $\sqrt{10\frac{1}{4}} - \frac{1}{2} - \sqrt{\sqrt{10\frac{1}{4}} - \frac{1}{2}}$.[4]

Problem II

Find two numbers the sum of which is the same as their product and the sum of the squares of which is 20. If their sum is x, their product is also x. Divide x into two parts the product of which is x, according to the rules given in the chapter on operations in the sixth book or according to II, 5 of the *Elements*. These parts will be $\frac{1}{2}x + \sqrt{\frac{1}{4}x^2 - x}$ and $\frac{1}{2}x - \sqrt{\frac{1}{4}x^2 - x}$, the sum of the squares of which is 20. Therefore since, as in the preceding, we have the relation of *binomium* and *recisum*, it suffices to square the parts of one of these and to multiply by two. Hence we will have, as the sum of the squares,

$$x^2 - 2x = 20,$$

wherefore x will be $\sqrt{21} + 1$. Likewise we do with these parts what was proposed and they will be $\sqrt{5\frac{1}{4}} + \frac{1}{2} + \sqrt{4\frac{1}{2} - \sqrt{5\frac{1}{4}}}$ and $\sqrt{5\frac{1}{4}} + \frac{1}{2} - \sqrt{4\frac{1}{2} - \sqrt{5\frac{1}{4}}}$.

Problem III

Find two numbers whose product equals their sum and the squares of which plus the numbers themselves make 20. Do as in the preceding and you will have as the sum $\sqrt{20\frac{1}{4}} + \frac{1}{2}$, which is 5. Since the squares of the parts plus the numbers themselves have to equal 20, therefore the squares alone, without the parts, will be 15. Hence divide 5 into two parts the sum of the squares of which is 15, and you will have these

---

(*cf.* the title of this chapter as given in the table of contents) and in the 14th problem in the *Ars Magna Arithmeticae* (*cf.* the reference to this problem in problem 31 of the same book.)

[3] 1570 and 1663 have "fourth."

[4] The text omits the internal radical signs.

numbers: $2\frac{1}{2}$[5] $+ \sqrt{1\frac{1}{4}}$ and $2\frac{1}{2} - \sqrt{1\frac{1}{4}}$.[6] Remember, however, that in the first operation, when you get $x^2 - 2x$ as the sum of the squares, you add x, which is the sum of the numbers, and [thus] arrive at

$$x^2 - x = 20.$$

Problem IIII

Find two numbers such that their product is their sum and such that, if 12 is divided by each of them, the sum of the squares of the quotients plus the sum of the dividends will be 80. (This and the preceding are Brother Luca's in a certain writing that has been lost.) Let the sum be x. Divide this into [two] parts [7]the product of which is x and you will have the parts[7] as you see them. Divide 12 by these, as in the figure:

Parts: $$\frac{12}{\frac{1}{2}x + \sqrt{\frac{1}{4}x^2 - x}} \qquad \frac{12}{\frac{1}{2}x - \sqrt{\frac{1}{4}x^2 - x}}$$

Squares: $$\frac{144}{\frac{1}{2}x^2 - x + \sqrt{\frac{1}{4}x^4 - x^3}} \qquad \frac{144}{\frac{1}{2}x^2 - x - \sqrt{\frac{1}{4}x^4 - x^3}}$$ [8]

Sum of the squares: $$\frac{144x^2 - 288x}{x^2}$$

Therefore, having squared these parts and added them, as I taught you in the fifth book, you will have the sum of the squares. To this add the sum of the dividends. Since, indeed, this is x,

$$\frac{144x^2 - 288x}{x^2} + x = 80.$$

Now multiply all terms by x, making

$$144x - 288 + x^2 = 80x.$$

Hence

$$x^2 + 64x^{[9]} = 288,$$

wherefore x equals $\sqrt{1312} - 32$. Divide $\sqrt{1312} - 32$ into two parts producing $\sqrt{1312} - 32$ and these will be the numbers sought for.

[5] 1545 has $2\frac{2}{2}$.

[6] 1663 omits the radical sign.

[7] 1570 and 1663 omit this passage.

[8] The text has

$$\frac{144}{\frac{1}{2}x^2 - x + \sqrt{\frac{1}{4}x^4 - x^3}}.$$

[9] 1570 and 1663 have $94x$.

Problem V

Find two numbers the sum of the squares of which is 20 and the product of which is equal to the square of their difference. This much is clear: These numbers have to be in proportion and this means that there must be a mean and two extremes. (It is also possible to solve this by the rule of equal position, for many problems can be solved by several diverse rules. Still we will do this according to the present rule.) Let x be the sum. We divide this into two parts the sum of the squares of which is 20 and these are as you see[10]: $\frac{1}{2}x + \sqrt{10 - \frac{1}{4}x^2}$ and $\frac{1}{2}x - \sqrt{10 - \frac{1}{4}x^2}$. Hence the square of their difference is $40 - x^2$, and this is equal to the product of the parts, which is $\frac{1}{2}x^2 - 10$. Therefore $1\frac{1}{2}x^2$ equals 50 and x equals $\sqrt{33\frac{1}{3}}$. Divide this into two parts the sum of the squares of which is 20, and these will be $\sqrt{8\frac{1}{3}} + \sqrt{1\frac{2}{3}}$ and $\sqrt{8\frac{1}{3}} - \sqrt{1\frac{2}{3}}$.[11]

And from this rule eight problems can be deduced which I have called, because of their very strong likeness,[12] the Sisters. [They fit] better with these cases than any others.

These eight problems, which are called the Sisters, follow. The last alone will serve as an example for the others.

Problem VI

Find two numbers the sum of whose squares is 10 and the sum of whose cubes is 30. Let x be the sum of the numbers. Divide it into parts such that the sum of their squares is 10. Then add the cubes of these parts and you will have

$$x^3 + 60 = 30x.$$

Problem VII

Find two numbers the sum of the squares of which is 10 and the difference between whose cubes is 15. Let their sum, as before, be x and you will have

$$x^6 = 300x^2 + 1100.$$

Problem VIII

Find two numbers the sum of the squares of which is 10 and the sum of the products of each of which and the square of the other is 20. Let

[10] 1545 and 1570 have *et erunt ut vides*; 1663 has *et erunt vi vides.*

[11] 1570 and 1663 have $\sqrt{\frac{2}{3}}$.

[12] *ob vehementum simulitudinem.*

x be the sum of the numbers in the same manner as before and you will have

$$x^6 + 300x^2 + 800x = 40x^4 + 1600.$$[13]

Problem IX

Find two numbers the sum of whose squares is 10 and the difference between the products of one of which and the square of the other is 4. Postulate, as before, x as the sum. You will then have

$$x^6 + 500x^2 = 40x^4 + 1936.$$

Problem X

Find two numbers the difference between the squares of which is 10 and the sum of whose cubes is 100. Let x be the sum of the numbers and, from this, work out parts whose squares differ by 10 and raise these to the cube and you will have

$$x^4 + 300 = 400x.$$

Problem XI

Find two numbers the difference between whose squares is 10, and the difference between whose cubes is 100. Postulate, as before, x as the sum of the numbers and you will have

$$x^4 + 33\tfrac{1}{3} = 13\tfrac{1}{3}x^3.$$

Problem XII

Find two numbers the difference between whose squares is 10 and the sum of the products of the square of one by the other is 100. Let x be the sum of the numbers as before and you will then have

$$x^4 = 400x + 100.$$

Problem XIII

Find two numbers the difference between the squares of which is 10 and the difference between the products of one of which and the square of the other is 100. This I will explain carefully as it [illustrates] the form of operation for the rest and is an example for them, not only the seven preceding but many others as well which can be formed in the same way. I assume, then, that x is the sum [of the numbers] and, in accordance with the rule of method or the chapter on operations in

[13] It is far from clear how this is derived from the premises.

the sixth[14] book, I divide it into two parts the difference between the squares of which is 10. This is [done] by dividing this difference — i.e., 10 — by twice that which is to be divided, that is, $2x$. The quotient is $5/x$. This is added to or subtracted from one-half that which is to be divided, which is $\frac{1}{2}x$. You will then have the parts and their squares. Place the latter under their roots but in reverse order, as thus shown:

$$\begin{array}{ll} \frac{1}{2}x + 5/x & \frac{1}{2}x - 5/x \\ \frac{1}{4}x^2 + 25/x^2 - 5 & \frac{1}{4}x^2 + 25/x^2 + 5. \end{array}$$

Multiply the lower by the upper. (In those cases in which we are looking for a difference, it is enough to multiply dissimilar parts — that is, what in one case produces a plus in the other produces a minus — just as in cases where they are to be added it suffices to multiply similar parts, for the remaining terms drop out of themselves.) Multiply therefore $\frac{1}{2}x$ by -5 and $5/x$ by $\frac{1}{4}x^2 + 25/x^2$.[15] Since where one produces a plus the other produces a minus, subtract the negative from the positive. This is the same as doubling one of the products. You will then have the difference between the products: $2\frac{1}{2}x - 250/x^3$. This equals 100. Divide through by $2\frac{1}{2}$ and multiply by x^3 and you will have

$$x^4 = 40x^3 + 100,$$

and similarly in the others.

You may derive from this [example] a rule of method by saying: Since two numbers the difference between the squares of which is given must produce a certain difference between the products of the multiplication of each by the square of the other,[16] x^4 will be equal to the square of the difference of the squares plus as many cubes as is that number which results from dividing the difference between the products by one-fourth the difference between the squares.[17] Thus, if I should say, Find two numbers the difference between the squares of which is 6 and the difference between the products of one and the square of the other is 60, we can say that

$$x^4 = 40x^3 + 36.$$

So, also, for the others.

[14] 1570 and 1663 have "fourth."

[15] 1570 and 1663 have, "Multiply $\frac{1}{2}x$ by $(-5/x - 5)$ and $(\frac{1}{4}x^2 + 5/x^2)$."

[16] *ex multiplicatione vicissim in quadrata.*

[17] I.e., if $a + b = x$, and $a^2 - b^2 = M$, $x^4 = M^2 + \left(\frac{a^2b - ab^2}{M/4}\right)x^3$.

RULE II

2. There is another form for the rule of the sum, far more subtle than the preceding, and it solves two unknowns and [performs] two conversions at one time. There is nothing more subtle among these rules. I discovered it in a certain fragment of Brother Luca's and have reduced it [to its present form] after many labors, since a picture of this rule could scarcely be gathered or perceived there. I will explain it readily. There is no method more general than this as far as showing the value of an unknown goes: Although it [pertains] most fully to the unknown, nothing else can be devised that is clearer.[18] Both example and rule will emerge[19] in the problems.

PROBLEM XIIII

Find two numbers the product of which is 8 and the sum of the squares of which plus the numbers themselves is 40. Let the sum of the numbers be $\frac{1}{2}q$, and let either of them be p. The other, therefore, is $\frac{1}{2}q - p$. Multiplied by each other they make $\frac{1}{2}pq - p^2$, and this is equal to 8. You have, accordingly,

$$p^2 + 8 = \tfrac{1}{2}pq.$$

Now follow the rule: Take one-half the coefficient of p, which is $\frac{1}{4}q$, as you were shown in the fifth chapter when the square and the number equal the first power. Then square $\frac{1}{4}q$, making $\frac{1}{16}q^2$, and subtract 8, the constant of the equation, making $\frac{1}{16}q^2 - 8$. Take the square root of this quantity and add it to or subtract it from $\frac{1}{4}q$, one-half the coefficient of p, and this gives the value of p, or the numbers you were seeking. One of these is $\frac{1}{4}q + \sqrt{\frac{1}{16}q^2 - 8}$, and the other is $\frac{1}{4}q - \sqrt{\frac{1}{16}q^2 - 8}$. Now add the squares of these to the sum of the numbers — i.e., to $\frac{1}{2}q$ — and this equals 40. Since the cross-products of $\frac{1}{4}q$ and $\sqrt{\frac{1}{16}q^2 - 8}$ are equal, one being plus, the other minus, they cancel out. The squares of the parts, therefore, are $\frac{1}{8}q^2 - 8$, and $\frac{1}{8}q^2 - 8$. Add, therefore, $\frac{1}{4}q^2 - 16$ and $\frac{1}{2}q$, the sum of the numbers. This is equal to 40. Now let x equal q and you will have

$$\tfrac{1}{4}x^2 + \tfrac{1}{2}x = 56.$$

Hence

$$x^2 + 2x = 224,$$

wherefore x equals $\sqrt{225} - 1$, or 14, and so much is the value of q. But we assumed the sum to be $\frac{1}{2}q$. Therefore the sum of the numbers is

[18] 1545 and 1570 have *praestantius*; 1663 has *prae antius*.

[19] reading *exit* for *erit*.

7. Divide 7 into two parts, the product of which is 8, and they will be $3\frac{1}{2} + \sqrt{4\frac{1}{4}}$ and $3\frac{1}{2} - \sqrt{4\frac{1}{4}}$, the numbers sought, and the squares of these plus the numbers themselves are 40.

Note. — And if it be asked of what use this rule is, or to whom it is of greater help than the first, I would answer: The first rule of the sixth book may be lacking in some particular in practice. This one, however, will serve freely right up to the very end, thus proving that whatever is most elegant, even more than that which is most useful, needs no outside help. Here is another example:

Problem XV

Find two numbers the product of which is 6 and the sum of the cubes of which is 100. Let $\frac{1}{2}q$ be the sum and let p[20] be one part. The other will be $\frac{1}{2}q - p$. Multiply the parts and you will have

$$\tfrac{1}{2}qp - p^2 = 6.$$

Follow out the equation as if $\frac{1}{2}q$ were a coefficient and you will have the solution. There are two values for p, namely $\frac{1}{4}q + \sqrt{\frac{1}{16}q^2 - 6}$,[21] and $\frac{1}{4}q - \sqrt{\frac{1}{16}q^2 - 6}$. The cubes of these must equal 100. Cube them, therefore, dropping the parts which are positive in one case and negative in the other, and you will have $\frac{1}{16}q^3 - 4\frac{1}{2}q$ for each of them. Hence, added together

$$\tfrac{1}{8}q^3 - 9q = 100.^{22}$$

Change q^3 to x^3 and q to x, and raise the coefficient of the cube to 1, and you will have

$$x^3 = 72x + 800$$

and the value of x will be the value of q, namely $\sqrt[3]{400 + \sqrt{146{,}176}} + \sqrt[3]{400 - \sqrt{146{,}176}}$.[23] One-half of this, which is $\sqrt[3]{50 + \sqrt{2284}} + \sqrt[3]{50 - \sqrt{2284}}$[24] is the sum of the numbers sought for and the parts are the cube roots of these, but this will appear in another operation.

[20] Here and the next place it occurs, the text has *res* instead of *positio*, i.e., x instead of p. I have translated as p, nevertheless, in order to avoid confusion with the *res* which occurs later on in the problem.

[21] 1570 and 1663 have 9.

[22] 1663 has 200.

[23] 1663 omits the second universal root.

[24] 1570 and 1663 have $\sqrt{2287}$.

Problem XVI

Find two numbers the difference between the squares of which is 10 and the greater of which plus the two squares is 40. Let x be the sum of the numbers and let $\frac{1}{2}y$ be one of them. The other will be $x - \frac{1}{2}y$. Square the parts and you will have $\frac{1}{4}y^2$ and $x^2 + \frac{1}{4}y^2 - xy$. Take the difference [between these], which is $xy - x^2$, and this is equal to 10. Therefore the value of x is $\frac{1}{2}y + \sqrt{\frac{1}{4}y^2 - 10}$ or $\frac{1}{2}y - \sqrt{\frac{1}{4}y^2 - 10}$, either of which is equal to x. Now x was divided into $\frac{1}{2}y$ and $x - \frac{1}{2}y$. Since $\frac{1}{2}y$ is common to both of these, it will be

$$\sqrt{\tfrac{1}{4}y^2 - 10} = x - \tfrac{1}{2}y.$$

Therefore the squares of the parts, which are $\frac{1}{4}y^2$ and $\frac{1}{4}y^2 - 10$, plus one of the parts — namely $\frac{1}{2}y$ — are equal to 40. Hence

$$y^2 + y = 100.$$

The unknown, therefore, which is y, equals $\sqrt{100\frac{1}{4}} - \frac{1}{2}$ and, since we assumed [one part as] $\frac{1}{2}y$, this will be $\sqrt{25\frac{1}{16}} - \frac{1}{4}$, that is, one-half of $\sqrt{100\frac{1}{4}} - \frac{1}{2}$, and the smaller one will be $\sqrt{15\frac{1}{8} - \sqrt{6\frac{17}{64}}}$. And generally in [using] this rule he who has the greater ingenuity will be the better in performance, for it takes many forms and much might be said of all of them. These, however, must suffice and we turn again to an example of the first rule, seeking by this method [the answer to] —

Problem XVII

Find two numbers the square of the second of which is equal to the product of the first and the sum [of the two], and the sum of the squares of which is 10. You see clearly that if x is postulated as the sum of them, it will be divided according to a proportion having a mean and two extremes and the parts will be $\sqrt{\frac{5}{4}x^2} - \frac{1}{2}x$ and $\frac{1}{2}x - \sqrt{\frac{5}{4}x^2}$.[25] The [sum of the] squares of these will therefore be $5x^2 - \sqrt{20x^4}$ and this will equal 10. Therefore, from the rule for adding plusses and minuses,[26]

$$5x^2 - 10 = \sqrt{20x^4}.$$

Hence the parts will be $\sqrt{2\frac{1}{2} + \sqrt{5}} + \sqrt{2\frac{1}{2} - \sqrt{5}}$ and $\sqrt{2\frac{1}{2} + \sqrt{5}} - \sqrt{2\frac{1}{2} - \sqrt{5}}$.

[25] 1545 has $\sqrt{\frac{5}{4}x^2 - \frac{1}{2}x}$ and $1\frac{1}{2}x - \sqrt{\frac{5}{4}x^2}$, using for the last the universal rather than the simple radical sign; 1570 and 1663 have the same except that $\sqrt{\frac{1}{4}x^2}$ appears as the last term instead of $\sqrt{\frac{5}{4}x^2}$.

[26] The text has *ex capitulo argumentandi p: et m:*. I read *augmentandi* in place of *argumentandi*.

Problem XVIII

Find three proportional numbers the first and second of which equal the third and the sum of the squares of the first and second of which is 10. Let the third be x. Divide x into two parts, the sum of the squares of which is 10. These will be $\frac{1}{2}x + \sqrt{5 - \frac{1}{4}x^2}$ and $\frac{1}{2}x - \sqrt{5 - \frac{1}{4}x^2}$. Multiply x by the smaller and the product is the square of the greater. Otherwise divide x according to a proportion having a mean and two extremes, then square the parts, and the sum of the squares will be 10. Hence the parts will be

$$\text{1st} \quad \sqrt{22\tfrac{1}{2} + \sqrt{405}} - \sqrt{12\tfrac{1}{2} + \sqrt{125}}$$

$$\text{2d} \quad \sqrt{12\tfrac{1}{2} + \sqrt{125}} - \sqrt{2\tfrac{1}{2} + \sqrt{5}}$$

$$\text{3d} \quad \sqrt{10 + \sqrt{80}}.$$

Problem XIX

Likewise, if someone should say, Find three numbers in proportion the product of the first and second of which is 10 and the first and second of which equal the third, by the same method of proceeding you will have the quantities:

$$\text{1st} \quad \sqrt{\sqrt{31\tfrac{1}{4}} + 5} - \sqrt{\sqrt{31\tfrac{1}{4}} - 5}$$

$$\text{2d} \quad \sqrt{\sqrt{31\tfrac{1}{4}} + 5} + \sqrt{\sqrt{31\tfrac{1}{4}} - 5}$$

$$\text{3d} \quad \sqrt{\sqrt{500} + 20}.$$

CHAPTER XXXVI

On the Rule of Free Position[1]

This is a rule for problems which follow from universal properties of numbers of which man is not aware and in seeking to solve which by other rules he labors hopelessly, for either they [i.e., the other rules] do not yield a proportion or they do not lead to a solution in all powers.[2] Such are these.

Problem I

Find five quantities the square of the second of which is equal to their sum plus the square of the first, and let the five quantities be in continued proportion. I can assume, therefore, whatever proportion I wish, beginning with x, as you see in the figure. In double [proportion], for example $4x^2$ will be the square of the second, and this will be equal to x^2, the square of the first, plus $31x$. Hence $3x^2$ equals $31x$, and x will be $10\frac{1}{3}$. The others [will follow] in double proportion as you see: $10\frac{1}{3}$, $20\frac{2}{3}$, $41\frac{1}{3}$, $82\frac{2}{3}$, and $165\frac{1}{3}$.

x^2	x
$4x^2$	$2x$
	$4x$
	$8x$
	$16x$

$$3x^2 = 31x$$

Problem II

Find two numbers in double proportion, and let [the sum of their] squares or cubes or fifth powers be equal to the [sum of the] numbers themselves. Let the example be of the fifth powers, since it is so much the more wonderful. We assume, therefore, x and $2x$, in double proportion, the sum of the fifth powers of which — $32x^5$ and x^5 — is $33x^5$.

[1] *De regula liberae positionis.* See footnote 1, page 192. In the 6th problem in the *Ars Magna Arithmeticae*, Cardano outlines the origin of the name of this rule thus: *Nota quod haec quaestio dicitur libertatis, quia verificatur in omni proportione, et ideo positio libera est. . . . Quando igitur homo ponit tibi rem difficilem, experiaris semper primo hanc regulam libertatis, quia forte in omni positione eveniet quod putas accidere in una tantum.*

[2] *non enim proportionem exigunt* [*exibunt?*], *nec tamen in omnibus quantitatibus queunt.*

This equals $3x$. Hence, according to the simple rule, x equals $\sqrt[4]{\frac{1}{11}}$, 3 having been divided by 33. The second quantity, therefore, will be $\sqrt[4]{1\frac{5}{11}}$, that is, twice $\sqrt[4]{\frac{1}{11}}$.

Problem III

Find three proportional numbers and let the proportion be triple and let the square of one-fourth their sum be one-seventh the second quantity. We assume, the quantities, therefore, to be x, $3x$, and $9x$, the sum of which is $13x$. One-fourth of this is $3\frac{1}{4}x$ and the square of this is $10\frac{9}{16}$ [x^2]. This is one-seventh of $3x$. Therefore $73\frac{15}{16}x^2$ equals $3x$, and x equals $\frac{48}{1183}$. The second quantity will be $\frac{144}{1183}$, and the third will be $\frac{432}{1183}$.

Problem IIII

Find three proportional numbers the second of which is 10, and one-twentieth the square of the sum of which is seven times the second. Let x be the first. The third, therefore, will be $100/x$. Since the square of one-twentieth the sum produces seven times the second, it produces 70, and $\sqrt{70}$ is one-twentieth the whole. Hence the whole is $\sqrt{28{,}000}$, and therefore [the sum of] the first and third will be $\sqrt{28{,}000} - 10$, and this is equal to $x + 100/x$. Hence

$$x^2 + 100 = x(\sqrt{28{,}000} - 10).$$

The first quantity, then, is $\sqrt{7000} - 5 - \sqrt{6925 - \sqrt{700{,}000}}$ and the third will be $\sqrt{7000} - 5 + \sqrt{6925 - \sqrt{700{,}000}}$. This could also be done more briefly without the unknown.

CHAPTER XXXVII

On the Rule for Postulating a Negative

RULE I

This rule is threefold, for one either assumes a negative, or seeks a negative square root, or seeks what is not. First, therefore, we will look for solutions for problems which can at least be verified[1] in the positive, as if someone should say

$$x^2 = 4x + 32$$

and, with the same solution,

$$x^2 = x + 20.$$

Now if you wished to go after a true solution, x would equal 8 in the first [equation], but in the second it would be 5. But if they are turned around, you say that

$$x^2 + 4x = 32$$

and that x equals 4; this will also be true for

$$x^2 + x = 20.$$

Say, therefore, that if $+4$ solves both these problems, -4 is the solution for

$$x^2 = 4x + 32$$

as well as for

$$x^2 = x + 20.$$

Hence you change the equations as we pointed out in the first chapter. If the case is impossible [to solve] with either a positive or a negative, the problem is a false one. If it is a true [problem] with a positive

[1] 1545 has *verificari minime possunt*; 1570 has *verare minime licet*; 1663 has *vera re minime licet.*

solution in one, it will be a true [problem] with a negative solution in the other. The following problem is of this sort:

Problem I

The dowry of Francis' wife is 100 *aurei* more than Francis' own property, and the square of the dowry is 400 more than the square of his property. Find the dowry and the property.

We assume that Francis has $-x$; therefore the dowry of his wife is $100 - x$. Square the parts, making x^2 and $10{,}000 + x^2 - 200x$. The difference between these is 400 *aurei*. Therefore

$$x^2 + 400 + 200x = 10{,}000 + x^2.$$

Subtract the common terms and you will have 9600 equal to $200x$, wherefore x equals 48 and so much he has in the negative — i.e., is lacking — and the dowry will be the residue of 100, namely 52. Therefore Francis has -48 *aurei*, without any capital or property, and the dowry of his wife is 52 *aurei*. By working this way, you can solve the most difficult and inextricable problems. The [next] is another example of this method:

Problem II

I have 12 *aurei* more than Francis, and the cube of mine is 1161 more than the cube of Francis'. Let $-x$ be Francis'. I have, then, 12 *aurei* minus x. Cube the parts, producing $-x^3$ and $1728+36x^2-432x-x^3$, and the difference between these is 1161. Hence

$$-x^3 + 432x\,[2] + 1161 = 1728 + 36x^2 - x^3.$$

Subtract $-x^3 + 1161$ from both sides, making

$$432x = 36x^2 + 567.$$

Therefore

$$x^2 + 15\tfrac{3}{4} = 12x,$$

wherefore x equals $1\frac{1}{2}$ and this, in the negative, is what Francis has, and I have $10\frac{1}{2}$, and such are the *aurei* sought for.

Problem III

Likewise, if I say I have 12 *aurei* more than Francis and the square of mine is 128 more than the cube of Francis' *aurei*, we let Francis have

[2] 1570 and 1663 have $422x$.

$-x$ and I have 12 *aurei* minus x. The square of mine is $144 + x^2 - 24x$, and this is equal to $-x^3 + 128$. Therefore

$$16 + x^2 + x^3 = 24x,$$

wherefore x is 4, and so much in the negative is what Francis lacks. I have 8 *aurei* of my own.

Rule II

The second species of negative assumption involves the square root of a negative. I will give an example: If it should be said, Divide 10 into two parts the product of which is 30 or 40, it is clear that this case is impossible. Nevertheless, we will work thus: We divide 10 into two equal parts, making each 5. These we square, making 25. Subtract 40, if you will, from the 25 thus produced, as I showed you in the chapter on operations in the sixth[3] book, leaving a remainder of -15, the square root of which added to or subtracted from 5 gives parts the product of which is 40. These will be $5 + \sqrt{-15}$ and $5 - \sqrt{-15}$.

Demonstration

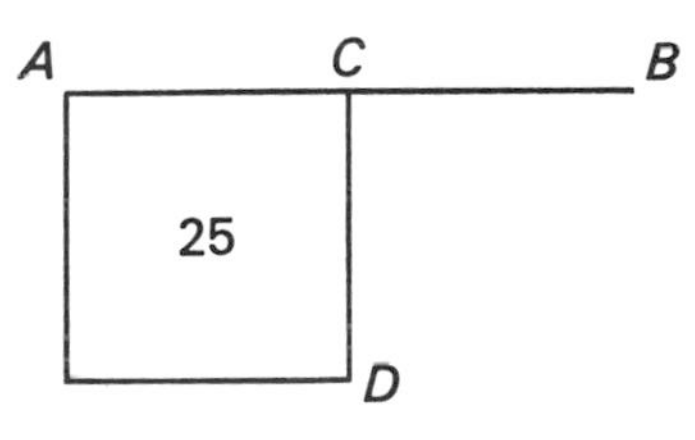

In order that a true understanding of this rule may appear, let AB be a line which we will say is 10 and which is divided in two parts, the rectangle based on which must be 40. Forty, however, is four times 10; wherefore we wish to quadruple the whole of AB. Now let AD be the square of AC, one-half of AB, and from AD subtract $4AB$, ignoring the number.[4] The square root of the remainder, then —if anything remains— added to or subtracted from AC shows the parts. But since such a remainder is negative, you will have to imagine $\sqrt{-15}$ —that is, the difference between AD and $4AB$— which you add to or subtract from AC, and you will have that which you seek, namely $5 + \sqrt{25 - 40}$ and $5 - \sqrt{25 - 40}$, or $5 + \sqrt{-15}$ and $5 - \sqrt{-15}$. Putting aside the mental tortures involved,[5] multiply

[3] 1570 and 1663 have "fourth."

[4] *absque numero.*

[5] *dimissis incruciationibus.* We may perhaps suspect Cardano of indulging in a play on words here, for this can also be translated "the cross-multiples having canceled out," with the whole sentence then reading, "Multiply $5 + \sqrt{-15}$ by $5 - \sqrt{-15}$ and, the cross-multiples having canceled out, the result is $25 - (-15)$, which is $+15$." *Cf.* the translation of this passage by Professor Vera Sanford in David Eugene Smith, *A Source Book in Mathematics* (New York, 1959 reprint), I, 202.

$5 + \sqrt{-15}$ by $5 - \sqrt{-15}$, making $25 - (-15)$ which is $+15$. Hence this product is 40. Yet the nature of *AD* is not the same as that of 40 or of *AB*, since a surface is far from the nature of a number and from that of a line, though somewhat closer to the latter. This truly is sophisticated,[6] since with it one cannot carry out the operations one can in the case of a pure negative and other [numbers]. [Likewise,] one cannot determine what it [the solution] is by adding the square of one-half the [given] number to the number to be produced and to or from the square root of this sum adding and subtracting half that which is to be divided.[7] For example, in this case you could divide 10 into two parts whose product is 40; add 25, the square of one-half of 10, to 40, making 65; from the square root of this subtract 5 and also add 5 to it; you then have parts with the likeness of $\sqrt{65} + 5$ and $\sqrt{65} - 5$. [But] while these numbers differ by 10, their sum is $\sqrt{260}$, not 10. So progresses arithmetic subtlety the end of which, as is said, is as refined as it is useless.

Problem IIII

Divide 6[8] into two parts, the sum of the squares of which is 50. This is solved by the first, not the second, rule for it involves a pure negative. Hence, square 3, one-half of 6, making 9. Subtract this from one-half of 50, which is 25, leaving 16 as the remainder. The square root of this is 4. Add this to and subtract it from 3, one-half of 6, making the parts 7 and -1, the sum of the squares of which is 50 and the sum of which is 6.

Problem V

This problem is solved by the same [rule]: Divide 6 into two parts the product of which is -40. Square 3, one-half of 6, making 9. Add this to 40, making 49, the square root of which is 7. This is added to 3, one-half of 6, and subtracted from the same, and you will have 10 and -4, which yield -40 when multiplied and 6 when added.

[6] In the 38th problem of the *Ars Magna Arithmeticae,* Cardano remarks: "Note that $\sqrt{9}$ is either $+3$ or -3, for a plus [times a plus] or a minus times a minus yields a plus. Therefore $\sqrt{-9}$ is neither $+3$ nor -3 but is some recondite third sort of thing" (*quaedam tertia natura abscondita*).

[7] *quae vere est sophistica, quoniam per eam, non ut in puro m: nec in aliis, operationes exercere licet, nec venari quid sit* [1663 at this point puts a period and inserts *Modus*] *est ut addas quadratum medietatis numeri numero producendo, et a* ℞ *aggregati minuas ac addas dimidium dividendi.*

[8] 1570 and 1663 have 8.

Likewise -10 and $+4$ produce -40 when multiplied and -6 when added. Hence this problem is one of a pure negative and belongs to the first rule.

Corollary.[9] From this it is evident that if it be said, Divide 6 into two parts the product of which is 40, the problem is one of the sophistic negative and pertains to the second rule. But if it is said, Divide 6 into two parts the product of which is -40, or divide -6 into two parts producing -40, in either case the problem will be one of the pure negative and will pertain to the first rule and the parts will be those that have been given. If it be said, Divide -6 into two parts the product of which is $+24$,[10] the problem will be one of the sophistic negative and will pertain to the second rule, and the parts will be $-3 + \sqrt{-15}$ and $-3 - \sqrt{-15}$.

Rule III

We are now able to pursue another species of negative which is neither a pure negative nor the square root of a negative but [in which] x is entirely false. This rule is composed as if of both [the others] and I give one example for it which is this:

Problem VI

Find three proportional numbers the square root of the first of which subtracted from the first gives the second and the square root of the second subtracted from the second gives the third. We assume, therefore, x^2 as the first. The second will be $x^2 - x$ and the third will be $x^2 - x - \sqrt{x^2 - x}$. Multiply the first by the third and the second by itself and, by doing so, you will have these quantities: $\frac{1}{4}$, $-\frac{1}{4}$, and $(-\frac{1}{4} - \sqrt{-\frac{1}{4}})$. The product of the first and third is $-\frac{1}{16} + \sqrt{\frac{1}{64}}$, which is $\frac{1}{8} - \frac{1}{16}$ and so much is produced by squaring the second number.

[9] In 1570 and 1663 the printer has misplaced this word and put it at the beginning of the next paragraph.

[10] The text has $+40$.

CHAPTER XXXVIII

How Parts and Powers Are Eliminated by Multiplication

Although [the matter covered by this chapter] has been generally and abundantly demonstrated in the fourth and fifth books,[1] it will be repeated here for the sake of ease and usefulness. This [the elimination of parts and powers] can be done in two ways, which we point out in as many rules. The first is particular and was discovered in connection with those four-term cases which were later demonstrated by us, as above, by geometrical reasoning. Now, having solved these, its usefulness is gone for the most part. Nevertheless, we will set it down for the enrichment of this art and the admiration of its ingenuity since, indeed, it may be useful for other [purposes] to which it can be applied quite easily even though it cannot be adapted to general use. This, therefore, is the rule: If you wish [two] different numbers the sum of the square of one of which and the cube of the other is a [given] number, divide their difference, using the unknown, into two parts, three times the square of one of which is equal to twice the other. Then, having found the parts, let x plus the part three times the square of which was taken to be the part to be cubed, and let x minus the part twice which was taken be the part to be squared.[2] Then, having completed the operation, you arrive at the cube and square equal to the number, the first power having dropped out.

Problem I

For example, find two numbers the difference between which is 8 and the sum of the cube of one of which and the square of the other is 100. First divide [the difference], by means of an unknown, into two parts three times the square of one of which is equal to twice the other.

[1] 1570 and 1663 have "third and fourth books."

[2] I.e., if $a^2 + b^3 = n$ and $a - b = m$, let $m = y + z$, y and z being such that $3y^2 = 2z$. Then $a = x - z$ and $b = x + y$.

These you will find to be 2 and 6, for three times 4, the square of 2, is 12, which is twice 6, the other number. Hence let the part to be cubed be $x + 2$ and the part to be squared $x - 6$. Add $(x + 2)^3$ and $(x - 6)^2$ and you have

$$x^3 + 7x^2 + 44 = 100.$$

Therefore

$$x^3 + 7x^2 = 56,$$

wherefore the value of x is $\sqrt[3]{15\frac{8}{27} + \sqrt{72\frac{16}{27}}} + \sqrt[3]{15\frac{8}{27} - \sqrt{72\frac{16}{27}}} - 2\frac{1}{3}$. Since the parts were $x + 2$ and $x - 6$, add 2 to this [value of x] and subtract 6 from it and you will have these parts: $\sqrt[3]{15\frac{8}{27} + \sqrt{72\frac{16}{27}}} + \sqrt[3]{15\frac{8}{27} - \sqrt{72\frac{16}{27}}} - \frac{1}{3}$ and $\sqrt[3]{15\frac{8}{27} + \sqrt{72\frac{16}{27}}} + \sqrt[3]{15\frac{8}{27} - \sqrt{72\frac{16}{27}}} - 8\frac{1}{3}$. It is evident, however, that one of these is a pure negative, and if you wanted both to be positive it would be necessary to assume that the cube and the square of the numbers were equal to a much larger number, perhaps 1000 instead of 100.

And you can use the same method if you wish the cube and square of [two] numbers differing by any given amount to differ by any assigned number. You find the parts of the difference by the same rule [as before]. Having found them, postulate in an opposite manner [from the preceding], namely x minus the number three times the square of which is to be taken and x plus the number twice which is to be taken. Then follow the operation as in the example.

Problem II

Find two numbers the difference between which is 8 and the difference between the cube of one of which and the square of the other is 100. Divide 8 into two parts as has been said, and they will be 2 and 6. Then postulate $x - 2$ and $x + 6$. Cube $x - 2$ and square $x + 6$, and take the difference. You will have

$$x^3 - 7x^2 - 44 = 100.$$

Therefore

$$x^3 = 7x^2 + 144,$$

wherefore the value of x is $\sqrt[3]{84\frac{19}{27} + \sqrt{7013\frac{1}{3}}} + \sqrt[3]{84\frac{19}{27} - \sqrt{7013\frac{1}{3}}} + 2\frac{1}{3}$. And since we assumed the parts are $x - 2$ and $x + 6$, the numbers sought for will be, as you see,

$$\sqrt[3]{84\frac{19}{27} + \sqrt{7013\frac{1}{3}}} + \sqrt[3]{84\frac{19}{27} - \sqrt{7013\frac{1}{3}}} + \frac{1}{3}$$

and

$$\sqrt[3]{84\frac{19}{27} + \sqrt{7013\frac{1}{3}}} + \sqrt[3]{84\frac{19}{27} - \sqrt{7013\frac{1}{3}}} + 8\frac{1}{3}.$$

And likewise, if it be said, Divide any number into two parts the sum of the square of one of which and the cube of the other is any [given] number, make two parts of the number to be divided, as above, to one of which — namely, the one three times the square of which is to be taken — add x, and from the other — namely, the one twice which is to be taken — subtract x. Then finish the operation, as in the example.

Problem III

Divide 8 into two parts the cube of one of which plus the square of the other is 400. Divide 8 into two parts as before, which will be 6 and 2, and assume $2 + x$ and $6 - x$. Cube $2 + x$ and square $6 - x$. Adding [the results], you will have

$$x^3 + 7x^2 + 44 = 400.$$

Therefore

$$x^3 + 7x^2 = 356,$$

wherefore the value of x is

$$\sqrt[3]{165\tfrac{8}{27} + \sqrt{27{,}161\tfrac{13}{27}}} + \sqrt[3]{165\tfrac{8}{27} - \sqrt{27{,}161\tfrac{13}{27}}} - 2\tfrac{1}{3}.$$

Hence, since the parts are $2 + x$ and $6 - x$, these will be, as you see,

$$8\tfrac{1}{3} - \sqrt[3]{165\tfrac{8}{27} + \sqrt{27{,}161\tfrac{13}{27}}} - \sqrt[3]{165\tfrac{8}{27} - \sqrt{27{,}161\tfrac{13}{27}}}$$

[and]

$$\sqrt[3]{165\tfrac{8}{27} + \sqrt{27{,}161\tfrac{13}{27}}} + \sqrt[3]{165\tfrac{8}{27} - \sqrt{27{,}161\tfrac{13}{27}}} - \tfrac{1}{3}.$$

And if it be said of the division of a given number into two parts that the difference between the cube of one and the square of the other is equal to [another] given number, then always assume $\frac{1}{3} + x$ as the part to be cubed and the remainder of the number to be divided — [i.e., the number] after $\frac{1}{3}$ is subtracted — minus x as the part to be squared. Then, having subtracted [one from the other], you will have a cube and first power equal to a number, wherefore both parts will be known speedily.

Problem IIII

For example, divide 8 into two parts the cube of one of which is greater than the square of the other by 10. We assume, then, the first part as $\frac{1}{3}$ and the other as $7\frac{2}{3}$, and we add x to $\frac{1}{3}$, making $\frac{1}{3} + x$, and subtract x from $7\frac{2}{3}$, making $7\frac{2}{3} - x$. From there we follow out the operation, and we will have $x^3 + x^2 + \frac{1}{3}x + \frac{1}{27}$ as the cube of $\frac{1}{3} + x$ and $x^2 - 15\frac{1}{3}x + 58\frac{7}{9}$ as the square. The difference between these is $x^3 + 15\frac{2}{3}x - 58\frac{20}{27}$, and this is equal to 10. Therefore

$$x^3 + 15\tfrac{2}{3}x = 68\tfrac{20}{27},$$

and the value of x is known. To this we add $\frac{1}{3}$ for the first part and from it we subtract $7\frac{2}{3}$ for the second. And if we had wished the square to be greater than the cube, we would have subtracted 10, the constant of the equation, from $58\frac{20}{27}$, and we would then have had

$$x^3 + 15\tfrac{2}{3}x = 48\tfrac{20}{27}.$$

The forms of this first rule are innumerable and are, as it were, part of the rule of method.

Rule II

This is another and easier rule that is much in use among us. It is such that it, also, will be easily understood from the remaining examples.

Problem V

Divide 8 into two such parts that, having taken the squares of both of them, as well as their cubes, and having multiplied one sum by the other, a perfect number is produced. I may say that it will also produce a determinate number[3] such as 10,000 or some other. The greatest that it can produce is 32,768, which is produced by the cube of the whole times the square of the whole, and the least it can produce is 4096. It must be determined, first, whether a perfect number falls between these two numbers. This is 8128. If there was none, the problem would be impossible. Assume now that one part is $4 - x$ and the other $4 + x$. The squares of these are $16 + 8x + x^2$ and $16 - 8x + x^2$ which, added together, make $32 + 2x^2$, the first powers having canceled out. The cubes, likewise, will be $64 + 12x^2 + 48x + x^3$ and $64 + 12x^2 - 48x - x^3$ which, added together, are $128 + 24x^2$. Hence we multiply $32 + 2x^2$ by $128 + 24x^2$ and get $4096 + 1024x^2 + 48x^4$, and these are equal to 8128. Hence we will have, having made the [proper] subtraction and division,

$$x^4 + 21\tfrac{1}{3}x^2 = 84,$$

wherefore x equals $\sqrt{\sqrt{197\frac{7}{9}} - 10\frac{2}{3}}$. Hence the parts are 4 plus the aforesaid root and 4 minus the same.

[3] *numerum terminatum.*

Problem VI

Divide 10 into two parts the cubes of the square roots of which are 26. Let [the sum of] these roots be x. Divide x into two parts, the sum of the squares of which is 10. Since the roots of these parts have to total x, you will have, according to the rules of the sixth book or according to Euclid, these parts: $\frac{1}{2}x + \sqrt{5 - \frac{1}{4}x^2}$ and $\frac{1}{2}x - \sqrt{5 - \frac{1}{4}x^2}$. These are to be cubed and since in cubing a *binomium* it is necessary to square each part, triple it, add to it the square of the other part, and multiply the sum[4] by that other part and, since these sums are similar and equal but one is plus and the other minus when we multiply three times the square of the first part plus the square of the second by the second, therefore it is sufficient to multiply the first part, which is $\frac{1}{2}x$, by three times the square of the second part, which is $15 - \frac{3}{4}x^2$, plus the square of the first part, which is $\frac{1}{4}x^2$, [which sum] is $15 - \frac{1}{2}x^2$.[5] But since, because of the two parts, this is a double operation, we will have $15 - \frac{1}{2}x^2$ multiplied by x, which is twice the $\frac{1}{2}x$ of the first part. This produces

$$15x - \tfrac{1}{2}x^3 = 26.$$

Hence

$$x^3 + 52 = 30x,$$

wherefore, according to the applicable rule, x equals $\sqrt{27} - 1$. Then you will have the parts that you see: $\sqrt{6\frac{3}{4}} - \frac{1}{2} + \sqrt{\sqrt{6\frac{3}{4}} - 2}$ and $\sqrt{6\frac{3}{4}} - \frac{1}{2} - \sqrt{\sqrt{6\frac{3}{4}} - 2}$. In verifying the operation, you need much more than this rule for ease. This we spoke of in the third book.

Problem VII

The following problem reduces to this [rule]: Someone bought one pound of saffron, two pounds of cinnamon, and five pounds of pepper, with proportional prices such that the price of all the pepper was to the price of the cinnamon as the price of the cinnamon was to the price of the saffron, the price of the saffron being the least, that of the pepper greatest, and that of the cinnamon in the middle, and the sum of these three prices was six *aurei*. Again, at the same prices, he bought 30 pounds of saffron, 50 pounds of cinnamon, and 40 pounds of pepper for 100 *aurei*. We want to know the price [per pound] of each.

[4] *productum.*

[5] 1570 and 1663 have $15 - \frac{3}{4}x^2$.

This problem was posed by Brother Luca,[6] but [as a problem] in proportional numbers, and he thought it extremely difficult, but it is not, for since the prices of five pounds of pepper, two of cinnamon, and one of saffron are proportionals, they remain proportionals also in larger quantities.[7] Hence we divide the 30 pounds of saffron by 1 — i.e., the second quantity by the first — and likewise the 50 pounds of cinnamon by 2, and the 40 of pepper by 5, and from these result the numbers given in the margin: that is, 30 for the saffron, 25 for the cinnamon, and 8 for the pepper. It is evident, therefore, that here are the numbers of three proportional quantities, which are the prices of one pound of saffron, two of cinnamon, and five of pepper, and that the first quantity or price taken 30 times, the second 25 times, and third 8 times, make 100 *aurei*. And truly these quantities, as was said, taken simply in themselves, are six *aurei*. Hence divide 6 into three proportional quantities, the first of which multiplied by 30, the second by 25, and the third by 8, make 100. We let the mean, therefore, be $2x$. The others, then, are $3 - x + \sqrt{9 - 3x^2 - 6x}$ and $3 - x - \sqrt{9 - 3x^2 - 6x}$. Each, then, is to be multiplied by its proper number. Since the first parts [i.e., $3 - x$] of the *binomia* are equal and both are positive, it is all the same whether they are multiplied by 30 and 8 or by 38. Likewise, since one of the universal radicals is negative and multiplied by 30 and the other is positive and multiplied by 8, the result will be the same, since they are equal, if [the negative radical] is multiplied by 22, the difference between 30 and 8. The parts produced will be $114 - 38x$ and $-\sqrt{4356 - 1452x^2}$[8] $- 2904x$, and these will be equal to 100. Add and subtract the like [terms][9] and you will have $14 + 12x$ equal to the universal root, which is negative, and therefore the square [of one equal] to the square [of the other]:

S.	*C.*	*P.*	*Aurei*
30	50	40	100
1	2	5	6
30	25	8	100

$$196 + 336x + 144x^2 = 4356 - 1452x^2 - 2904x.$$

[6] A similar but not identical problem is given by Luca Paccioli on p. 94r of his *Sũma de Arithmetica* (*supra*, note 4, p. 8). It is written in terms of two pounds each of saffron, cinnamon, and mastic for 20 ducats, the mastic being to the cinnamon as the cinnamon is to the saffron. In a later transaction eight pounds of saffron, 12 of cinnamon, and 108 of mastic are bought for 200 ducats. The question is, of course, the price per pound for each. Luca says the problem was put to him in Florence on 22 June 1480.

[7] *in suis aggregatis.*

[8] The text has $\sqrt{1452x^2}$.

[9] This includes, of course, the addition of $50x$ for the cinnamon.

These are equal parts and you will then have

$$4160 = 1596x^2 + 3240x.$$

Therefore

$$x^2 + 2\tfrac{4}{133}[x] = 2\tfrac{242}{399},$$

and the value of x is $\sqrt{3\frac{33,794}{53,067}} - 1\frac{2}{133}$.[10] Hence the price of one pound of saffron is $4\frac{2}{133} - \sqrt{3\frac{33,794}{53,067}} - \sqrt{1\frac{4595}{53,067} + \frac{12}{133}\sqrt{3\frac{33,794}{53,067}}}$ *aurei*,[11] the price of two pounds of cinnamon is $2\sqrt{3\frac{33,794}{53,067}} - 2\frac{4}{133}$ *aurei*,[12] and the price of five pounds of pepper is

$$4\tfrac{2}{133} - \sqrt{3\tfrac{33,794}{53,067}} + \sqrt{1\tfrac{4595}{53,067} + \tfrac{12}{133}\sqrt{3\tfrac{33,794}{53,067}}}$$

aurei.[13] If, therefore, you divide these proportional prices by their respective number of pounds — i.e., the first by 1, the second by 2, and the third by 5 — you will have the price of each pound of each kind, and if you multiply these by the two numbers in the second purchase — the price of saffron by 30, that of the cinnamon by 50, and that of the pepper by 40 — you will have the amount of money each cost.

Problem VIII

This problem is solved by the same method: Divide 14 into three proportional parts, the greatest of which multiplied by 2, the middle one by 3, and the smallest by 4, make 36 when the products are added together. You arrive by the method used above at

$$x^2 + 9\tfrac{1}{3}x = 53\tfrac{1}{3},$$

wherefore x equals $\sqrt{75\frac{1}{9}} - 4\frac{2}{3}$, which is 4. This is the mean quantity, for the mean quantity was assumed to be x, not $2x$ as before.

Problem IX

Divide 14 into three such proportional quantities that the sum of twice the first and three times the second equals seven times the third.

[10] The text has $\sqrt{3\frac{4,494,602}{7,057,911}} - 1\frac{2}{133}$, a less simplified form of what is given above.

[11] The text has $4\frac{2}{133} - \sqrt{3\frac{4,494,602}{7,057,911}}$ *aurei*.

[12] The text uses the form $\sqrt{14\frac{3,862,586}{7,057,911}} - 2\frac{4}{133}$.

[13] The text has $\sqrt{3\frac{4,494,602}{7,057,911}} + 1\frac{131}{133}$.

Let the second be $2x$. The others are, as you see, $7-x+\sqrt{49-14x-3x^2}$ and $7-x-\sqrt{49-14x-3x^2}$. Multiplying the second by 3 makes $6x$. In the same way, the first is to be multiplied by 2 and the third by 7, and the [proper] subtractions are to be made. Therefore, since the parts of both are alike and since the first [half[14]] of both is positive while the second is positive in the first and negative in the third, it suffices to multiply the first by the difference between 7 and 2, which is 5, and this produces $35 - 5x$ for the [first half] of the third part. Having subtracted from this $6x$, the product of [three times] the second part, we will have $35 - 11x$ as the difference between the second and third products. The first [half of the third part,] however, is produced by multiplying 9, the sum of the first and third [multipliers] by the universal root, and this makes $\sqrt{3969 - 1134x - 243x^2}$. This, therefore, equals $35 - 11x$, and the square [of one equals] the square [of the other]. Hence

$$1225 - 770x + 121x^2 = 3969 - 1134x - 243x^2.$$

[Having subtracted] like parts, you will have

$$2744 = 364x + 364x^2,$$

wherefore

$$x^2 + x = 7\tfrac{7}{13}.$$

Hence the value of x is known[15] and twice this is the second part, namely $\sqrt{31\frac{2}{13}}$[16] $- 1$.

Problem X

Divide 8 into three proportional quantities such that the sum of the squares of the first and third[17] will be three times the square of the second. Let the middle quantity by $2x$. The square of this is $4x^2$, three times which is the sum of the squares of the first and third. The first, moreover, is $4 - x + \sqrt{16 - 8x - 3x^2}$ and the third is $4 - x - \sqrt{16 - 8x - 3x^2}$. To square these, therefore, you see that it is necessary to multiply the universal root by itself once and the first part by itself once, and both [terms] are positive; therefore it is sufficient

[14] I.e., $(7 - x)$.

[15] It is $\sqrt{7\frac{41}{52}} - \frac{1}{2}$.

[16] The text has $\sqrt{31\frac{2}{3}}$.

[17] The text has "second."

to double these products. Then the universal root must be multiplied by twice the first part. Wherefore, since in one case a positive answer results and in the other a negative, the parts being equal, the result is zero. Therefore we will have as the sum of the squares $64 - 32x - 4x^2$ and this is equal to $12x^2$, three times the square of the second. Therefore

$$x^2 + 2x = 4,$$

wherefore x equals $\sqrt{5} - 1$, and twice this is the middle quantity, namely $\sqrt{20} - 2$, and the others are, as you see, $5 - \sqrt{5} + \sqrt{6 - \sqrt{20}}$ and $5 - \sqrt{5} - \sqrt{6 - \sqrt{20}}$. The square of the second is $24 - \sqrt{320}$ and the squares of the first and third have proved to be $72 - \sqrt{2880}$.[18]

But if it were said that the squares of the first and third are [equal to] three times the squares of the second and third,[19] the solution would be derived through this rule only with difficulty. Certainly it would be far easier through the first rule of the 39th chapter. By postulating the quantities as 1, x, and x^2 you will have $x^4 + 3(x^2 + 1)$.[20] Hence x would be known.

Problem XI

If you say, Divide 8 into two parts the sum of the quotients of each of which when divided by the square of the other is 10, let the parts be $4 + x$ and $4 - x$, and following the rule you come to a derivative equation of the fourth power, square, and constant. This is an easy one.

Problem XII

Find four numbers in continued proportion, the sum of the first, second, and fourth of which is 15, and of the first, third, and fourth is 17. You say that, since these sums differ by the difference between the second and third, the third is two more than the second. Hence I let the second be $x - 1$ and the third $x + 1$, for thus their difference will

[18] 1570 and 1663 have $\sqrt{2580}$.

[19] *Sed si dicerat, quod quadrata primae et tertiae, tripla essent quadratis secundae et tertiae.* Except for the fact that this is being contrasted with the principal problem, one would be tempted to translate this as "three times the squares of the first and third are [equal to] the squares of the second and third."

[20] *habebis 1 quadrati quadratum p: triplum de quadrato p: 1.* This does not seem to match the statement of the problem as given in the preceding note. Either there is missing material here or there is superfluous material there.

be 2. The sum of the first and fourth, then, is $16 - x$. Multiply the second by the third, producing $x^2 - 1$. Divide $16 - x$ into two parts, the mutual product of which is $x^2 - 1$, and these parts will be $8 - \frac{1}{2}x + \sqrt{65 - 8x - \frac{3}{4}x^2}$ and $8 - \frac{1}{2}x - \sqrt{65 - 8x - \frac{3}{4}x^2}$. Since the ratio of the fourth to the third is that of the second to the first, from the very nature [of a continued proportion], and since the product of the second and third is equal to the product of the first and fourth, it suffices, in order to demonstrate that they are in continued proportion, that the sum of the cubes of the second and third is equal to the products of the first and fourth times their mutual squares. And such cubes derive from the multiplication of three times the square of their second part [that is, 1] plus the square of the first part [that is, x^2] by this same first part, because the other product — three times the square of the first part plus the second part by this same second — drops out since what is positive in one case is negative in the other. Therefore we will have as the sum of the cubes $2x^3 + 6x$ and so much should [also] result from the mutual multiplication of the squares of the first and fourth quantities by the quantities themselves. This, however, as has been demonstrated, is equivalent to multiplying the product of the two quantities by the sum of the same quantities, according to the sixth book. Multiply, therefore, the quantities by each other. Since the universal roots are the same, the cross multiplication [of the two parts of the *binomia*, $(8 - \frac{1}{2}x)$ and $\sqrt{65 - 8x - \frac{3}{4}x^2}$] will produce nothing. Hence it suffices to square each of these parts and to subtract one from the other, since a minus times a plus produces a minus. Therefore the products of these similar parts will be $x^2 - 1$. But the sum of these quantities[21] is $16 - x$, because the universal roots drop out. Therefore the product [of the sum of the quantities and their own product] will be $16x^2 - x^3 + x - 16$, and this is equal to $2x^3 + 6x$. Hence

$$3x^3 + 5x + 16 = 16x^2.$$

Hence x is [found] by the rule. You see, however, that since this inextricable problem is reducible with great ease, it can also be reduced to a rule of method, for where the difference is 2, $3x^3$ plus $5x$ plus a constant equidistant between the two sums will always be equal to x^2 with a coefficient equal to the constant.

[21] The text has "the sum of these roots."

Problem XIII

ABC is a right triangle and *AD* is perpendicular to its base. Its side, *AB*, plus *BD* is 36 and *AC* plus *CD* is 24. Find its area.

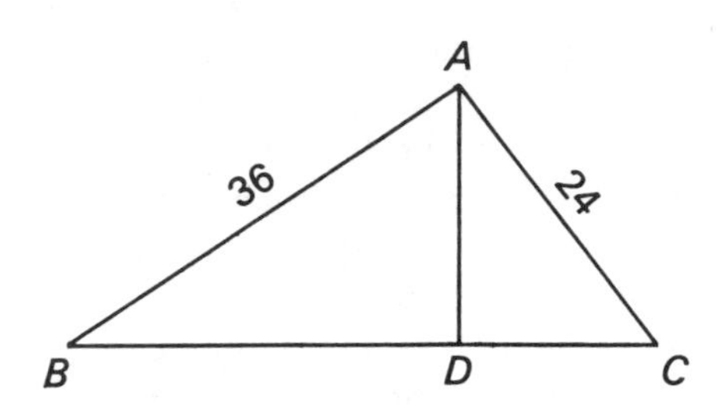

Let *BC* equal x. Therefore BC^2 will equal x^2. Hence, since

$$AB + BD = 36$$

and

$$AC + CD = 24,$$

the sum of its sides will be 60. Therefore

$$AB + AC^{22} = 60 - x.$$

Divide, then, $AB + AC$ into two parts the sum of the squares of which is equal to BC^2, according to the teaching of I, 47 of Euclid's *Elements*.[23] Hence divide $60 - x$ equally, according to the rules of our own sixth book, making $30 - \frac{1}{2}x$. Square this, making $900 - 30x + \frac{1}{4}x^2$. Subtract this from $\frac{1}{2}BC^2$ and the remainder is $\frac{1}{4}x^2 + 30x - 900$, the square root of which, added to or subtracted from $\frac{1}{2}(AB + AC)$, shows the parts. Hence *AB* equals $30 - \frac{1}{2}x + \sqrt{\frac{1}{4}x^2 + 30x - 900}$, and *AC* equals $30 - \frac{1}{2}x - \sqrt{\frac{1}{4}x^2 + 30x - 900}$. Therefore if *AB* is subtracted from $AB + BD$, *BD* is left as $6 + \frac{1}{2}x - \sqrt{\frac{1}{4}x^2 + 30x - 900}$, and likewise by subtracting *AC* from $AC + CD$, *CD* remains as $\frac{1}{2}x - 6 + \sqrt{\frac{1}{4}x^2 + 30x - 900}$. It is evident, however, from the demonstration in I, 47 of Euclid's *Elements*, that the difference between AB^2 and AC^2 is equal to the difference between BD^2 and CD^2. The difference, however, between two quantities is always in the dissimilar parts, for what are alike produce no difference. Therefore, since the squares of the parts consist of nine products of which three are the squares of the parts and these three are entirely similar in the comparison of *AB* to *AC* and of *BD* to *CD* and the two products of 30 and $\frac{1}{2}x$ are common in *AB* and *AC*, since both are negative, and the products of 6 and the universal square root are common in *BD* and *CD*, for both are negative, therefore the difference between [the squares of] *AB* and *AC* is the product of 30 and the universal square root in the case of *AB* and the product of $\frac{1}{2}x$ and the universal root in the case of *AC*.[24] Hence the difference between AB^2 and AC^2 is

[22] The text has *BC*.

[23] 1545 has *secundum doctrinam 47ᵉ primi elemen. Euclidis*; 1570 and 1663 have *per 47 primi Elementorum Euclidis.*

[24] *differentia igitur AB et AC, ex parte AB, est multiplicatio 30 in* ℞ *v: et ex parte AC, multiplicatio ½ positionis in* ℞ *v:.*

that [amount] by which $\sqrt{225x^2 + 27{,}000x - 810{,}000}$[25] exceeds $\sqrt{\frac{1}{16}x^4 + 7\frac{1}{2}x^3 - 225x^2}$. By the same reasoning,[26] the difference between BD^2 and CD^2 is that by which $3x$ is greater than

$$\sqrt{\tfrac{1}{16}x^4 + 7\tfrac{1}{2}x^3 - 225x^2}.$$

It would be right, however, in carrying out this operation, to multiply everything by four. But this we will not both with here, for if[27] one quadruplicate equals another quadruplicate, one simple thing will equal the other. These differences are supposed to be equal and the universal roots are also equal. Therefore, by common consent, $3x$ equals the first universal root, that is $\sqrt{225x^2 + 27{,}000x - 810{,}000}$, wherefore

$$216x^2 + 27{,}000x = 810{,}000,$$

and

$$x^2 + 125x = 3750,$$

wherefore x will equal $\sqrt{7656\frac{1}{4}} - 62\frac{1}{2}$, which is 25, and such is BC, from which you can derive the others.

Problem XIIII

Let, again, ABC be a right triangle and let AD be its perpendicular, and let

$$AB + CD = 29$$

and

$$AC + BD = 31.$$

The area is to be found.

Let BC be x, and AB and[28] AC will be the same as in the above problem. But beware that you do not place the greater side on the side of the greater number as before. Subtract, therefore, AB from 29 and AC from 31, and you will have these quantities:

$$\begin{array}{ll} AB & 30 - \frac{1}{2}x + \sqrt{\frac{1}{4}x^2 + 30x - 900} \\ AC & 30 - \frac{1}{2}x - \sqrt{\frac{1}{4}x^2 + 30x - 900} \\ BD & \frac{1}{2}x + 1 + \sqrt{\frac{1}{4}x^2 + 30x - 900} \\ CD & \frac{1}{2}x - 1 - \sqrt{\frac{1}{4}x^2 + 30x - 900}. \end{array}$$

[25] 1570 and 1663 have $2700x$.

[26] 1545 and 1570 have *eadem est ratione*; 1663 has *eadem ratione.*

[27] 1663 omits this word.

[28] 1570 and 1663 omit this word.

The difference, then, between AB^2 and AC^2 is equal to the difference between BD^2 and CD^2. The difference between AB^2 and AC^2, however, is as before, but the difference between BD^2 and CD^2 is, as you see, $2x$[29] + [2] $\sqrt{\frac{1}{16}x^4 + 7\frac{1}{2}x^3 - 225x^2}$, this having been [derived] by following the same method as in the preceding question, although the excess is absolute, not mutual as it was there. Since, then, AB^2 exceeds AC^2 by the difference between BD^2 and CD^2, the difference between BD^2 and CD^2 added to AC^2 will constitute AB^2. Therefore

$$\sqrt{225x^2 + 27{,}000x - 810{,}000} = \tfrac{1}{2}x + \sqrt{\tfrac{1}{4}x^4 + 30x^3 - 900x^2},$$

for these quantities are, respectively, one-fourth of the difference between BD^2 and CD^2 and one-fourth of that part of AB^2 by which it exceeds AC^2.[30] Therefore, squaring the parts, we will have

$$1125\tfrac{1}{4}x^2\,[31] + 27{,}000x - \tfrac{1}{4}x^4 - 30x^3 - 810{,}000$$
$$= \sqrt{225x^4 + 27{,}000x^3 - 810{,}000x^2}$$

and when you square these parts you will come to something which has no solution[32] and therefore must be solved by some special method.[33] I wanted you to see, however, how easy it is to work in this fashion and how very difficult the problem is unless it is solved with the help of geometry. For it is evident that BD equals 25 as in the previous problem. Truly this must be the general solution. The sides of the triangle, therefore, are these: BC, 25; AB, 20; AC, 15; AD, 12; BD, 16; CD, 9. Hence the area is 150.

[29] The text has $\frac{1}{2}x$.

[30] The text has "for this universal root is the sum of the universal root of the difference between BD^2 and CD^2 and that part of AC^2 by which AC^2 exceeds AB^2: *nam haec R̸ v: est aggregatum ex R̸ v: differentiae quadratorum BD et CD, et partis quadratis AC, in qua superat quadratum AB.*"

[31] The text has $675\frac{1}{4}x^2$.

[32] 1545 has *quae non habent aestimationem*; 1570 and 1663 have *cuius non est nota aestimatio.*

[33] 1545 has *et ideo solvenda est regula particulari*; 1570 and 1663 have *quare alia regula indigebis aut generali aut speciali.*

CHAPTER XXXIX

On the Rule by Which We Find an Unknown Quantity in Several Stages[1]

RULE I

This rule is similar to the rule for a mean. This is it: Set up as many terms in whatever powers you wish as the number of items to be found.[2] Then look for their ratio. Having found this, substitute it for the terms to be found, as proposed. Carry out the operation and you will have an equation and, having this, you will have the value of the unknown.

PROBLEM I

For example: Find three proportional quantities of which the square of the first is equal to the second and third, and the square of the third is equal to the squares of the first and second. Since, therefore, the square of the third is equal to the squares of the second and first, let

$$x^4 = x^2 + 1.$$

Therefore x, or the ratio, is $\sqrt{\sqrt{1\frac{1}{4}} + \frac{1}{2}}$. Hence we substitute 1 y, $y\sqrt{\sqrt{1\frac{1}{4}} + \frac{1}{2}}$, and $y(\sqrt{1\frac{1}{4}} + \frac{1}{2})$. The square of the first quantity, therefore, which is y^2, is equal to the second and third, namely to that many y's. Therefore the value of y is the sum of the second and third, since to divide anything by 1, which is the coefficient of y^2, is not to divide it. Therefore the value of y is $\sqrt{1\frac{1}{4}} + \frac{1}{2} + \sqrt{\sqrt{1\frac{1}{4}} + \frac{1}{2}}$, and the second quantity is the product of this and $\sqrt{\sqrt{1\frac{1}{4}} + \frac{1}{2}}$, and the third will be had by multiplying y, which you have, by $\sqrt{1\frac{1}{4}} + \frac{1}{2}$.[3]

[1] *De regula qua pluribus positionibus invenimus ignotam quantitatem.* This is also known as the "iterative rule"; see the title of this chapter as given in the table of contents. See also footnote 2, page 205.

[2] *Constitue quantitates totidem in denominationibus liberis, quotus est numerus quaerendarum.*

[3] 1545 has $\frac{2}{2}$.

Problem II

Find three proportional numbers the third of which is equal to the second and first, and the square of the first of which is equal to the sum of the second and third.

Let the first be 1, the second x, and the third x^2.[4] Since the third is equal to the second and first,

$$x^2 = x + 1,$$

and the ratio, accordingly, is $\sqrt{1\frac{1}{4}} + \frac{1}{2}$. Then the parts will be y, $y(\sqrt{1\frac{1}{4}} + \frac{1}{2})$, and $y(1\frac{1}{2} + \sqrt{1\frac{1}{4}})$. And since the square of the first is equal to the sum of the second and third, therefore

$$y^2 = y(\sqrt{1\tfrac{1}{4}} + \tfrac{1}{2} + 1\tfrac{1}{2} + \sqrt{1\tfrac{1}{4}}),$$

wherefore y will equal $\sqrt{5} + 2$, and the parts will be, as you see, $\sqrt{5} + 2$; $3\frac{1}{2} + \sqrt{11\frac{1}{4}}$; and $\sqrt{31\frac{1}{4}} + 5\frac{1}{2}$.

Problem III

Find four proportional quantities the square of the fourth of which is equal to the squares of the first and second and the sum of which is 10. I take 1, x, x^2, and x^3 [as the quantities]. Hence

$$x^6 = x^2 + 1.$$

Therefore, according to the rule for derivatives, x equals

$$\sqrt{\sqrt[3]{\tfrac{1}{2} + \sqrt{\tfrac{23}{108}}} + \sqrt[3]{\tfrac{1}{2} - \sqrt{\tfrac{23}{108}}}}.$$

Hence, 1 having been assumed to be the first quantity, this is the second, and the third will be its square, namely

$$\sqrt[3]{\tfrac{1}{2} + \sqrt{\tfrac{23}{108}}} + \sqrt[3]{\tfrac{1}{2} - \sqrt{\tfrac{23}{108}}},$$

and the fourth will be the cube of the second or ratio. Hence, having added the four quantities — namely 1, x, x^2, and x^3 — and having divided 10 by their sum, the first quantity will appear. This multiplied by x gives us the second, and multiplied by x again the third, and this multiplied by x will give us the fourth.

Problem IIII

Find four numbers in continued proportion, the square of the fourth of which is equal to the squares of the first and third, and the sum [of all four] of which is 10.

[4] The text has these three in reverse order.

As before, I take 1, x, x^2, and x^3 [as the quantities]. Therefore

$$x^6 = x^4 + 1.$$

Hence, according to the rule for derivatives, x equals

$$\sqrt{\sqrt[3]{\tfrac{29}{54} + \sqrt{\tfrac{31}{108}}} + \tfrac{1}{3} + \sqrt[3]{\tfrac{29}{54} - \sqrt{\tfrac{31}{108}}}},$$

and the square of this, which is the same with the universal radical sign removed, is the third quantity. Then, having multiplied the second and third by each other, or having cubed the second, or squared the third, and having added 1, the fourth arises.[5] Having added the four together, if you will divide 10 by them, you will have the first one sought for and, having multiplied this by the second, third, and fourth of the preceding, you will have the second, third, and fourth quantities which you are seeking.

Rule II

2. There is another rule, more noble than the preceding. It is Lodovico Ferrari's, who gave it to me on my request. Through it we have all the solutions[6] for equations of the fourth power, square, first power, and number, or of the fourth power, cube, square, and number, and I set them out here in order:

1. $x^4 = bx^2 + ax + N$
2. $x^4 = bx^2 + cx^3 + N$
3. $x^4 = cx^3 + N$
4. $x^4 = ax + N$
5. $x^4 + cx^3 = bx^2 + N$.[7]
6. $x^4 + ax = bx^2 + N$
7. $x^4 + cx^3 = N$
8. $x^4 + ax = N$
9. $x^4 + bx^2 = cx^3 + N$
10. $x^4 + bx^2 = ax + N$
11. $x^4 + bx^2 + ax = N$
12. $x^4 + bx^2 + cx^3 = N$
13. $x^4 + bx^2 + N = cx^3$
14. $x^4 + bx^2 + N = ax$
15. $x^4 + N = cx^3 + bx^2$
16. $x^4 + N = cx^3$
17. $x^4 + N = ax + bx^2$
18. $x^4 + N = ax$
19. $x^4 + cx^3 + N = bx^2$
20. $x^4 + ax + N = bx^2$

In all these cases, therefore, which are indeed [only] the most general as there are 67 others, it is convenient to reduce those involving

[5] So in the text, even though it cannot be so: *inde ductis invicem secunda et tertia, vel secunda ad suum cubum, vel tertia ad quadratum, et addita unitate consurgit quarta.*

[6] *omnes aestimationes ferme*, an expression which can mean either "absolutely all" or "nearly all the solutions."

[7] 1570 and 1663 have $x^4 = ax + N$, a repetition of the preceding.

the cube to equations in which x is present, as the seventh to the fourth or the second to the first. Now we look for the demonstration this way:

Demonstration

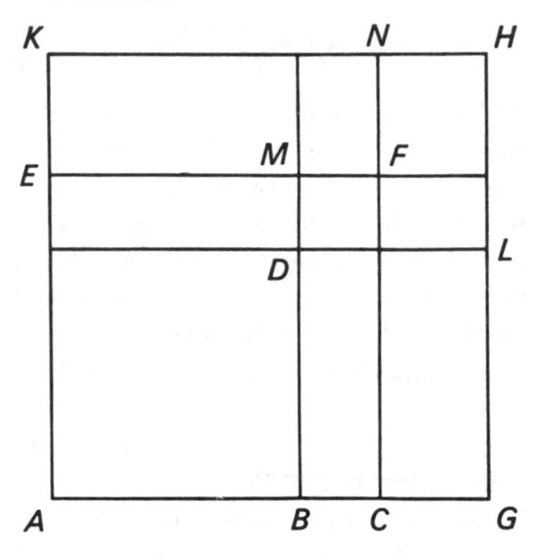

3. Let the square AF be divided into two squares, AD and DF, and two supplements, DC and DE, and let me add the gnomon KFG around it in order to complete the whole square AH. I say that this gnomon will consist of GC^2 plus twice the added line $GC \times CA$, for FG is $GC \times CF$, from the definition given at the beginning of the second book of the *Elements*, and CF equals CA by the definition of a square. Since, according to I, 43[8] of the *Elements*, KF equals FG, the two surfaces GF and FK consist of $GC \times 2CA$, and GC^2 equals FH, according to the corollary to II, 4 of the *Elements*. Hence the proposition is clear. If, therefore, AD equals x^4, and CD and DE [each] equal $3x^2$, and DF equals 9, BA will equal x^2 and BC will necessarily equal 3. Since, therefore, we shall wish to add more squares to DC and DE, these will be CL and KM. In order to complete the whole square, the surface LMN is necessary. This, as has been demonstrated, consists of the square of GC [plus $2GC \times BC$], one-half the [original] number of squares,[9] for CL is the surface produced by $GC \times AB$, as has been shown, and AB is x^2 because we assumed that AD is x^4 and, therefore, FL and MN are made up of $GC \times CB$, according to I, 42 of the *Elements*. Hence the surface LMN (this is the number to be added) is $GC \times 2CB$ (that is, times the coefficient of x^2, which is 6) plus GC times itself (that is, times the added number of squares). This demonstration is our own.

4. In carrying out this operation, always reduce the side of [the equation with] x^4 [in it] to its square root — that is, by adding so much to both sides that $x^4 + bx^2 + N$ will have a square root. This is

[8] The text has I, 44.

[9] I.e., BC is half the original number of squares.

easy, since you can assume that one-half the coefficient of x^2 is the root of the constant. Do this in such fashion that the highest terms in both solutions are positive.[10] Otherwise the *trinomium* or the *binomium* enlarged to a *trinomium* will necessarily lack a root.

5. Then, having done all this, add as many squares and such a number to one side, in accordance with the third rule, as have to be added to the other side — the side which has the first power — in order to make a *trinomium* having a square root with x. You will thus have a certain number of squares and a number to be added to both sides. Having gone this far, extract the square root of both. In one case it will be x^2 plus or minus a number and in the other one or more x's plus or minus a number or a number minus the x's. Then the solution can be derived from Chapter V of this [work].

Problem V

For example, divide 10 into three proportional parts, the product of the first and second of which is 6. This was proposed by Zuanne de Tonini da Coi,[11] who said it would not be solved. I said it could, though I did not yet know the method [for doing so]. This was discovered by Ferrari.

Let the mean be x. The first, then, will be $6/x$ and the third will be $\frac{1}{6}x^3$. These equal 10. Multiplying all terms by $6x$, we will have

$$60x = x^4 + 6x^2 + 36.$$

Following the fifth rule, add $6x^2$ to both sides and you will have

$$x^4 + 12x^2 + 36 = 6x^2 + 60x,\text{[12]}$$

for if equals are added to equals the sums are equal. Moreover, $x^4 + 12x^2 + 36$ has a square root which is $x^2 + 6$. Now if $6x^2 + 60x$

[10] *item facies, ut denominationes extremae sint plus, in ambabus aequationibus.* Perhaps, "Do this in such fashion that the highest terms on both [sides of the] equation are positive."

[11] *Fl.* 1530. Cardano, in his *De Libris Propriis* (*Opera Omnia*, I, 66), gives us this thumbnail sketch of da Coi and tells us how it was from him that he first learned that certain forms of the cubic had been solved: *Tunc vero contigit ut quidam Brixiensis nomine Ioannes Colla, vir procerae staturae, macilentus, pallidus, subniger, cavis oculis, mansuetis moribus, tardus in ambulando, rarus verbis, ingeniosus, ac in mathematicis exercitatus, Mediolanum veniret: retulitque inventas esse duas regulas Algebrae, ut vocant, cubi et numeri invicem comparatorum, sciscitatus sum a quo, a Scipione Ferreo Bononiensis, inquit: Quis habet? dixit, Nicolaus Tartalea et Antoninus Maria Floridus, sed Tartalea cum Mediolanum venisset, illas me docuit, quamvis satis invitus. Has cum diligenter perscrutatus essem cum Ludovico Ferrario inventa demonstratione alias innumeras etiam adinvenimus, ut ex his librum artis magnae conficerem.*

[12] 1570 and 1663 have $90x$.

[likewise] had a square root, we would have the solution. But it does not. Hence there must be added to both sides alike enough squares and a number so that on one side there is a *trinomium* with a root and on the other the same.[13] Let b, therefore, be a number of squares and since, as you see in the figure for the third rule,

$$CL + MK = 2GC \times AB,$$

and GC is b, I always let $2b$ — that is, $2GC$ — be the additional number of squares. And, since the number to be added to 36 is LMN, or $GC^2 + (2GC \times CB)$ or $(GC \times 2CB)$, [$2CB$] being 12, the prior coefficient of x^2, I multiply b, one-half the additional number of squares, by the prior coefficient of x^2 and by itself, making $b^2 + 12b$ to be added to the second side, and [I likewise take] $2b$ as the [additional] number of squares. We have, therefore, by common consent again, these quantities equal to each other:

$$x^4 +^{14} (2b + 12)x^2 + (b^2 + 12b) + 36 = (2b\,[+]\,6)x^2 + 60x + (b^2 + 12b).$$

Each of these has a square root, the first according to the third rule and the second by supposition. Hence multiplying the first part of the *trinomium* by the third gives the square of one-half the second part.[15] Since, therefore, the square of one-half the second part is $900x^2$ and the product of the first and third is $(2b^3 + 30b^2 + 72b)x^2$, dividing both by x^2 — since equals divided by equals produce equals — will yield

$$2b^3 + 30b^2 + 72b = 900,$$

wherefore

$$b^3 + 15b^2 + 36b = 450.$$

Reducing this to a rule: It is always sufficient to have b^3 plus one and a quarter times the coefficient of x^2 [for the coefficient of b^2] plus such a number of b's as was the original constant of the equation. Thus, if we had

$$x^4 + 12x^2 + 36 = 6x^2 + 60x,$$

we will have

$$b^3 + 15b^2 + 36b = 450,$$

[13] Reading *ibidem* for the *autem* of the text, which would otherwise be incomplete if not meaningless: *ut in priore relinquatur trinomium habens radicem, in altero autem fiat.*

[14] 1545 has *p*:; 1570 and 1663 have *po*.

[15] I.e., $(2b + 6)x^2(b^2 + 12b) = (30x)^2$.

[the constant being] one-half the square of one-half the coefficient of x. And if we had

$$x^4 + 16x^2 + 64 = 80x,$$

we will have

$$b^3 + 20b^2 + 64b^{16} = 800.$$

And if we had

$$x^4 + 20x^2 + 100 = 80x,$$

we will have

$$b^3 + 25b^2 + 100b = 800.$$

Hence, having what we had in the former example, we have

$$b^3 + 15b^2 + 36b = 450.$$

Therefore, according to Chapter XVII, b equals $\sqrt[3]{190 + \sqrt{33,903}} + \sqrt[3]{190 - \sqrt{33,903}} - 5$.[17] Twice this, therefore, is the number of squares which must be added to both sides, since we agreed $2b$ are to be added. The number to be added to both sides is, by the demonstration, the square of this plus the product of this and 12, the coefficient of x^2. It is evident, moreover, that the square root of the first sum, absent the additional amount,[18] is always x^2 plus one-half the coefficient of x^2 [19]or [x^2] plus b plus one-half the prior coefficient of x^2 [with the additional amount]. Thus

$$x^4 + 6x^2 + 9 = 144$$

and

$$x^4 + (2b + 6)x^2 + (b^2 + 6b) + 9 = 225$$

and the square root [of the latter] is $x^2 + b + 3$. Moreover $b + 3$ is half of $2b + 6$, the [new] coefficient of x^2, and therefore, since b is of a different nature from x^2, its value must first be found. This is the simple number that must be added. In the present example,[19] it will be

[16] 1570 and 1663 have $94b$.

[17] The text, here and at later points, has $\sqrt[3]{287\frac{1}{2} + \sqrt{80,449\frac{1}{4}}} + \sqrt[3]{287\frac{1}{2} - \sqrt{80,449\frac{1}{4}}} - 5$.

[18] *absque alio.*

[19] This material appears in 1570 and 1663 but not in 1545.

$\sqrt[3]{190+\sqrt{33,903}}+\sqrt[3]{190-\sqrt{33,903}}+1$. (This [i.e., $+1$] is because one-half the prior coefficient of x^2 is 6 and in the added *trinomium* [the integer] is -5. So the whole is, as I have said, a true [number].) The remaining part is x^2 [with a coefficient of] 6 plus twice this number. So the coefficient of x^2 is $\sqrt[3]{1520+\sqrt{2,169,792}}+\sqrt[3]{1520-\sqrt{2,169,792}}-4$;[20] the coefficient of x, by supposition, is 60; and the constant is (as has been shown) the square of the aforesaid quantity plus 12 times the same. Since, by supposition, the coefficient of x^2 times the constant of the equation is the square of one-half the coefficient of x, divide 900, the square of one-half the coefficient of x, by the coefficient of x^2, and the number will result. The quantities, then, are these:

[coefficient of] x^2: $\sqrt[3]{1520+\sqrt{2,169,792}}+\sqrt[3]{1520-\sqrt{2,169,792}}-4$
[coefficient of] x: 60

constant: $$\frac{900}{\sqrt[3]{1520+\sqrt{2,169,792}}+\sqrt[3]{1520-\sqrt{2,169,792}}-4}$$

Now since the side *AG* is composed of the sides of the two squares *AD* and *DH*, forgetting about the supplements, the sum of the square roots of the first and third of these quantities will be the square root of the whole sum. Therefore the square root of the first and third quantities equals $x^2+\sqrt[3]{190+\sqrt{33,903}}+\sqrt[3]{190-\sqrt{33,903}}+1$. But the square root of the first quantity is a number of x's, since it is the square root of this many squares, and the square root of the third quantity is a number since the third quantity is a number. We will have, accordingly, x^2 and a number equal to x and [another] number. Having taken its square root — i.e., having taken the square root of the denominator and the numerator — subtract the smaller number from the larger and you will have x^2 plus this whole number minus the number written hereafter — i.e.,

$$\frac{30}{\sqrt{\sqrt[3]{1520+\sqrt{2,169,792}}+\sqrt[3]{1520-\sqrt{2,169,792}}-4}}$$

[20] The text, here and at later points, has

$$\sqrt[3]{2300+\sqrt{5,148,752}}+\sqrt[3]{2300-\sqrt{5,148,752}}-4.$$

— equal to this number of x's, namely $\sqrt[3]{1520 + \sqrt{2,169,792}} + \sqrt[3]{1520 - \sqrt{2,169,792}} - 4$.[21] Nor does it matter that this number is composed of positives and negatives, for it is the same to say

$$x^2 + 8 = 6x$$

as to say

$$x^2 + 10 - 2 = 6x.$$

Follow, then, [the rule of] Chapter V on the square and number equal to the first power by multiplying one-half the coefficient of x by itself and subtracting the constant, then taking the square root of the remainder and adding to this one-half the coefficient of x. You will then have x, which is the mean of the proportional quantities you were seeking.

Problem VI

Find a number which is equal to its square root plus twice its cube root. You say, then, that if such a number is x^6, its square root is necessarily x^3 and twice its cube root is $2x^2$. Therefore

$$x^6 = x^3 + 2x^2.$$

Reducing these by x^2 to lower powers,

$$x^4 = x + 2.$$

(I posed twice the cube root since, while the rule is general, this can be solved in two ways, as will be seen.) Now if

$$x^4 = x + 2$$
$$x^4 - 1 = x + 1,$$

for equals have been subtracted from equals. Divide, therefore, both of these by $x + 1$ as a common divisor and you will have

$$x^3 - x^2 + x - 1 = 1,$$

wherefore

$$x^3 + x = x^2 + 2.$$

Therefore, according to Chapter XVIII, x equals $\sqrt[3]{\sqrt{\frac{2241}{2916}} + \frac{47}{54}} - \sqrt[3]{\sqrt{\frac{2241}{2916}} - \frac{47}{54}} + \frac{1}{3}$, and the sixth power of this is the number sought

[21] The typography in 1570 and 1663 is very mixed up at this point; inserted in the midst of this expression is a slew of other figures which, in 1545, are set off from the body of the text.

for and is equal to its square root plus twice its cube root, and these roots are twice the square of this quantity plus its cube.

But by the general rule we would work thus: Since

$$x^4 = x + 2,$$

we add to both sides $2bx^2$[22] which we write thus so that you may understand that this is not of the nature of the first terms but of the nature of the coefficient of x^2.[23] Hence the number to be added is the square of the coefficient of x^2, and this is, as in the third rule of this chapter, the square DF. This addition of the supplements [i.e., $2bx^2$] is that of DC and DE[24] to the simple square AD. Hence it suffices to add the square DF without adding the surfaces FL and MN which were necessary in the example of the fifth problem. Since, therefore, adding $2b[x^2] + b^2$ to $x + 2$ makes a total of $2b[x^2]$[25] $+ x + 2 + b^2$ and this has a root, it must be that the square of half the middle quantity, x, equals the product of the extremes. Therefore,

$$\tfrac{1}{4}x^2 = (2b^3 + 4b)x^2$$

and, having divided both sides by x^2,

$$\tfrac{1}{4} = 2b^3 + 4b$$

and

$$\tfrac{1}{8} = b^3 + 2b,$$

wherefore b equals $\sqrt[3]{\sqrt{\frac{2075}{6912}} + \frac{1}{16}} - \sqrt[3]{\sqrt{\frac{2075}{6912}} - \frac{1}{16}}$. Twice this, therefore, is the number of squares to be added to both sides and the square of this will be the number to be added to both sides and, for the sake of clearer understanding, I set them [the two sides] out here:

$$\text{I. } x^4 + x^2\left(\sqrt[3]{\sqrt{19\tfrac{23}{108}}^{\,26} + \tfrac{1}{2}} - \sqrt[3]{\sqrt{19\tfrac{23}{108}} - \tfrac{1}{2}}\right) + \sqrt[3]{\tfrac{1051}{3456} + \sqrt{\tfrac{2075}{442{,}368}}} + \sqrt[3]{\tfrac{1051}{3456} - \sqrt{\tfrac{2075}{442{,}368}}} - 1\tfrac{1}{3}.^{27}$$

[22] *2 positiones quadratorum.*

[23] *cui subscribimus qd. ut intellegas non esse ex genere priorum denominationum, sed esse positiones quadratorum.*

[24] 1545 has *DC, AC, DE*. I take *AC* to be a misprint for *ac* ("and"). 1570 and 1663 use lower-case letters the whole way through, with similar confusion.

[25] *2 positiones numeri quadratorum.*

[26] The text has $\frac{23}{2075}$ here and in the following places where the fraction is repeated, except as otherwise noted.

[27] 1663 has $\frac{1}{3}$.

$$\text{II. } x^2\left(\sqrt[3]{\sqrt{19\tfrac{23}{108}} + \tfrac{1}{2}} - \sqrt[3]{\sqrt{19\tfrac{23}{108}} - \tfrac{1}{2}}\right) + x$$
$$+ \sqrt[3]{\tfrac{1051}{3456}\text{[28]} + \sqrt{\tfrac{2075}{442,368}}} + \sqrt[3]{\tfrac{1051}{3456} - \sqrt{\tfrac{2075}{442,368}}} + \tfrac{2}{3}.$$

It is therefore clear that the square root of the first is

$$x^2 + \sqrt[3]{\sqrt{\tfrac{2075}{6912}} + \tfrac{1}{16}} - \sqrt[3]{\sqrt{\tfrac{2075}{6912}} - \tfrac{1}{16}}$$

and the square root of the second is

$$x\sqrt{\sqrt[3]{\sqrt{19\tfrac{23}{108}}\text{[29]} + \tfrac{1}{2}} - \sqrt[3]{\sqrt{19\tfrac{23}{108}} - \tfrac{1}{2}}}$$
$$+ \sqrt{\sqrt[3]{\tfrac{1051}{3456} + \sqrt{\tfrac{2075}{442,368}}} + \sqrt[3]{\tfrac{1051}{3456} - \sqrt{\tfrac{2075}{442,368}}} + \tfrac{2}{3}}$$

and this, as I said, is composed of the roots of the extremes, without bothering about the supplements,[30] just as the side of the square AF[31] is made up of AB and BC. It is evident, also, that the number which is on the side of x^2 is less than the number on the side of x. Therefore we will have

$$x^2 = x\left(\sqrt{\sqrt[3]{\sqrt{19\tfrac{23}{108}} + \tfrac{1}{2}} - \sqrt[3]{\sqrt{19\tfrac{23}{108}} - \tfrac{1}{2}}}\right)$$
$$+ \sqrt{\sqrt[3]{\tfrac{1051}{3456} + \sqrt{\tfrac{2075}{442,368}}} + \sqrt[3]{\tfrac{1051}{3456} - \sqrt{\tfrac{2075}{442,368}}} + \tfrac{2}{3}}$$
$$- \sqrt[3]{\sqrt{\tfrac{2075}{6912}} + \tfrac{1}{16}} - \sqrt[3]{\sqrt{\tfrac{2075}{6912}} - \tfrac{1}{16}}.$$

Hence we square one-half the coefficient of x — that is, we multiply the whole thing by itself, which is the same as removing the universal radical sign, and take one-fourth of the product, and this is one-half of the last universal root [and is the same as] the negative set out above. Therefore, having added, the entire number is left made up of

$$\sqrt{\sqrt[3]{\tfrac{1051}{3456} + \sqrt{\tfrac{2075}{442,368}}} + \sqrt[3]{\tfrac{1051}{3456} - \sqrt{\tfrac{2075}{442,368}}}}$$
$$\overline{+ \tfrac{2}{3} - \sqrt[3]{\sqrt{\tfrac{2075}{442,368}} + \tfrac{1}{128}} - \sqrt[3]{\sqrt{\tfrac{2075}{442,368}} - \tfrac{1}{128}}},$$

and the most universal root of all this plus one-half the coefficient of x, — that is, plus the number

$$\sqrt{\sqrt[3]{\sqrt{\tfrac{2075}{442,368}} + \tfrac{1}{128}} - \sqrt[3]{\sqrt{\tfrac{2075}{442,368}} - \tfrac{1}{128}}}$$

— is x.

[28] 1570 and 1663 have $\tfrac{1051}{3454}$.

[29] 1545 has $\tfrac{23}{2071}$.

[30] *lateribus quadratorum extremorum, absque commemoratione supplementorum.*

[31] The text has AD.

And if it had been said that a given number is equal to its square root plus its cube root, it would be insolvable according to the general rule except by this second method. However, I showed you in the book on irrational quantities how to reduce equal values to the same, as a first solution to the second, no matter how difficult the operation may be. And therefore carrying out such operations as these is about the greatest thing to which the perfection of human intellect or, rather, of human imagination, can come. In this you will recognize the difference between the two.

Problem VII

If, now, someone says, Find a number which multiplied by its own cube root plus 6 equals 64, you will say, Having assumed this number to be x^3 we will have

$$x^4 + 6x^3 = 64.$$

Hence, according to the seventh rule of Chapter VII for transforming [equations] we will have

$$x^4 = 6x + 4$$

whence, having derived the solution for this equation according to the ninth rule of the same chapter, we will have the desired [answer]. And to whomever these operations seem very difficult, as they may truly be believed to be, we will show another method by which these irrational quantities serving as numbers[32] can be reduced to numbers and, furthermore, we will give two demonstrations of it, one geometrical *a causa* and the other arithmetical *ab effectu.*

Problem VIII

Divide 6 into three proportional quantities the sum of the squares of the first and second of which is 4. Let the first be x. The square of this is x^2. The difference between this and 4, then, is the square of the second quantity, that is, $4 - x^2$. Subtract x plus the square root of this from 6, and you will have the third quantity, $6 - x - \sqrt{4 - x^2}$. Hence the product of the first and third will [equal] the square of the second:

$$6x - x^2 - \sqrt{4x^2 - x^4} = 4 - x^2.$$

[32] *quantitates istae irrationales aequivalentes numeris.*

Remove $-x^2$ from both sides and you will have

$$4 = 6x - \sqrt{4x^2 - x^4}$$

wherefore

$$6x - 4 = \sqrt{4x^2 - x^4},$$

the squares of which are also equal. From these subtract the common $4x^2$ from both sides, and you will then have

$$32x^2 + 16 + x^4 = 48x.$$

Hence by adding 240 to both sides — that is, the difference between the square of one-half the coefficient of x^2 [and the constant] — you will have

$$x^4 + 32x^2 + 256^{33} = 48x + 240.$$

Add, therefore, $2bx^2 + (b^2 + 32b)$ to both sides. The first part, therefore, necessarily has a root and since we wish to have one also for the second, which is $2bx^2 + 48x$ (this from the previous equation) + $(b^2 + 32b) + 240$, we multiply the first part of the *trinomium* by the third, as you see, and we square one-half the second, and this makes

$$576x^2 = (2b^3 + 64b^2 + 480b)x^2$$

wherefore

$$288 = b^3 + 32b^2 + 240b,$$

wherefore, by Chapter XVII of this [work], we will have

$$b^3 = 101\tfrac{1}{3}b + 420\tfrac{20}{27}.^{34}$$

Then, having derived the solution for this according to the proper rule, subtract $10\frac{2}{3}$, one-third the coefficient of b^2, in accordance with Chapter XVII, and there arises the solution for b.[35] You will then have x^2 plus 16 plus this solution on one side equal to x with a coefficient equal to the square root of twice the solution just discovered, plus the square root of the sum of the square of this solution and the same multiplied by 32, plus 240, the additional number. This, however, as is clear, is less than the former number since if 256 were added in place of 240, they would be equal. Hence x^2 plus the solution discovered plus 16 minus the universal square root of these three quantities — that is,

[33] 1570 and 1663 have 156.

[34] The text has $420\frac{16}{27}$.

[35] *et consurgit rei fictae aestimatio.*

the square of the solution plus 32 times the same, plus 240 as, so to speak, the number — is equal to x with a coefficient of the root of twice the solution discovered, which is what was proposed.

Problem IX

Find a number the fourth power of which plus four times itself plus 8 is equal to ten times its square. We will say, therefore,

$$x^4 + 4x + 8 = 10x^2.$$

Therefore, as always, we give the x's to the squares, taking them from x^4, and we will have

$$x^4 + 8 = 10x^2 - 4x$$

and, since we see that the coefficient of x^2 is great and that of x small, we will attempt to decrease the coefficient of x^2 rather than to augment it. This we will do by subtracting $2x^2$ from both sides although the usual practice would be, on the contrary, to begin [by subtracting] from the minor quantity [rather than] from the squares[36] since it is not proper that a negative second power should appear on the side of the first power, because there would then be no square root. Having, however, subtracting $2x^2$ from both sides, you will have

$$x^4 - 2x^2 + 8 = 8x^2 - 4x.$$

It is clear, however, that if $x^4 - 2x^2$ is to have a square root, the number must be $+1$. But it is $+8$. So 7 must be subtracted from both sides, and we will accordingly have

$$x^4 - 2x^2 + 1 = 8x^2 - 4x - 7.$$

We will therefore add in the negative, so to speak, $2bx^2$ to the remaining $-2x^2$, in accordance with the rule, and we will add $b^2 + 2b$ positively, as by the same rule, to the number on both sides. Therefore the two sides will be equal, for the things that were added and subtracted were equal. Hence $(8 - 2b)x^2 - 4x + (b^2 + 2b) - 7$ has a square root. By multiplying, therefore, the first part, which is $(8 - 2b)x^2$ by the third, which is $(b^2 + 2b) - 7$, what you see in the margin $[(8b^2 + 16b - 56 - 2b^3 - 4b^2 + 14b)x^2]$ results as the coefficient of x^2 and this must be equal to $4x^2$, which is the number produced by squaring one-half the middle part. Therefore, by canceling x^2 from

[36] *nam a minori imo a 2 quadratis semper ferme est incipiendum.* The *2* before *quadratis* appears to be an error.

both sides, the multinomial becomes equal to 4. Whence, reducing the parts [still further] to their likenesses,

$$2b^3 + 60 = 4b^2 + 30b$$

and

$$b^3 + 30 = 2b^2 + 15b,$$

wherefore b equals 2, whether working according to the rule or by one's own sense alone.

Note. There are three things to be noted about this: First, that I reduced b in this problem to a number so that you could see its true value more clearly, for it is always foolish to add difficulty to difficulty. Second, that

$$b^3 + 30 = 2b^2 + 15b$$

has another solution besides 2, which can be found by its proper rule, but in order that this long operation shall not detain us any longer we leave it for the present. Third, that you see that the demonstration holds as well for the negative as for the positive and that the number must always be added, since (whether the squares are added or subtracted) it arises from the square of [one-half] the number of [new] squares plus the prior coefficient of x^2 times one-half the number of squares [which, in this case, are] being subtracted.

This being so, we may say that the value of b is 2 and that $2bx^2$ is to be added negatively. Hence we subtract $4x^2$ from both sides and we will accordingly have

$$x^4 - 6x^2 + 1 = 4x^2 - 4x - 7.$$

For the number, however, there must be added the square of one-half the

$$\begin{array}{r} x^4 - 2x^2 + 1 \\ - 4x^2 + 8 \\ \hline x^4 - 6x^2 + 9^{1} \\ \hline 8x^2 - 4x - 7 \\ -4x^2 \quad\quad + 8 \\ \hline 4x^2 - 4x + 1 \end{array}$$

$$1x^4 - 6x^2 + 9 \quad \begin{cases} x^2 - 3 \\ 3 - x^2 \end{cases}$$

$$4x^2 - 4x^2 + 1 \quad \begin{cases} 2x - 1 \\ 1 - 2x \end{cases}$$

$$x^2 = 2x + 2 \qquad \sqrt{3} + 1$$

$$x^2 + 2x = 4 \qquad \sqrt{5} - 1$$

	First Solution	Second Solution
x	$\sqrt{3} + 1$	$\sqrt{5} - 1$
x^2	$4 + \sqrt{12}$	$6 - \sqrt{20}$
x^4	$28 + \sqrt{768}$	$56 - \sqrt{2880}$
$4x$	$\sqrt{48} + 4$	$\sqrt{80} - 4$
x^4	$\sqrt{768} + 28$	$-\sqrt{2880} + 56$
	$+ 8$	$+ 8$
Sum	$\sqrt{1200} + 40$	$60 - \sqrt{2000}$
$10x^2$	$40 + \sqrt{1200}$	$60 - \sqrt{2000}$

[1] The text has 1

[2] The text has $4x^2$

[coefficient of the] subtracted squares and this half is 2, the square of which is 4, and likewise the product of the prior coefficient of x^2 and the value of b, which product is 4. Therefore we add 8 to both sides, thus making, as you see

$$x^4 - 6x^2 + 9 = 4x^2 - 4x + 1.$$

It is evident, moreover, that both [sides] have double roots, as you see, but reduction having been made, they necessarily come to two equations:

$$\text{either} \quad x^2 = 2x + 2$$
$$\text{or} \quad x^2 + 2x = 4.$$

The solutions for these equations are $\sqrt{3} + 1$ and $\sqrt{5} - 1$. I say, therefore, that with these solutions

$$x^4 + 4x + 8 = 10x^2.$$

The proof of this is given clearly at the side, as is evident. I need not say whether, having found another value for b in the case of

$$b^3 + 30 = 2b^2 + 15b$$

we would come to two other solutions [for x]. If this operation delights you, you may go ahead and inquire into this for yourself.

Problem X

Find three proportional numbers the sum of which is 8 and the square of the third of which is equal to the sum of the squares of the first and second. According to the first rule, we let these be 1, x, and x^2. Their squares, therefore, will be 1, x^2, and x^4. Hence

$$x^4 = x^2 + 1.$$

Hence, according to Chapter XXIV on derivatives, we will have as the value of x $\sqrt{\sqrt{1\frac{1}{4}} + \frac{1}{2}}$ and of the third quantity — that is, the square of this — $\sqrt{1\frac{1}{4}} + \frac{1}{2}$, and the first was 1. Hence the whole sum is $1\frac{1}{2} + \sqrt{1\frac{1}{4}} + \sqrt{\sqrt{1\frac{1}{4}} + \frac{1}{2}}$. This, however, is not 8 as was proposed. Ask, therefore, what 1, the first quantity, would be according to the rule of three,[37] if $1\frac{1}{2} + \sqrt{1\frac{1}{4}} + \sqrt{\sqrt{1\frac{1}{4}} + \frac{1}{2}}$ were 8. Multiply 8 by 1, making 8. Divide 8 by $1\frac{1}{2} + \sqrt{1\frac{1}{4}} + \sqrt{\sqrt{1\frac{1}{4}} + \frac{1}{2}}$, and the result is $4 + \sqrt{\sqrt{500} + 10} - \sqrt{\sqrt{1620}}$[38] $+ 18$, and this is the first quantity.

[37] *per regulam trium quantitatum.*

[38] The text has $\sqrt{1920}$.

Having this, if you multiply it by $\sqrt{1\frac{1}{4}} + \frac{1}{2}$ you will have the third quantity and, if you again multiply the first quantity by the last one found, the square root of the entire product is the second quantity. Do not be astonished that I preferred the third quantity to the second in this operation, for this is by far the simpler [way to solve the problem].

Problem XI

If someone says, Find a number which multiplied by its own cube root minus 3 is 64, let the number be x^3. Hence multiplying this by its cube root minus 3 yields

$$x^4 - 3x^3 = 64.$$

I say that this can be solved in the manner of the seventh problem. It can also be solved by another method without a transformation, by which the seventh problem could also have been solved and more easily [than it was]. I wished, however, to teach both these methods so that you may know which works the more easily. You ought to know two things: First, just as x should always remain on the side on which the number and second power are and should not be on the side of x^4, so the cube, whether positive or negative, should remain with x^4. Secondly, just as the coefficient of x should not change, so also the coefficient of x^3 ought not to vary. To these we can also add a third, namely that [just as] where there are x's we will arrive at the fourth power plus the square plus the number equal to the second power plus or minus the x's plus a number, so in this case we will arrive at the fourth and second powers plus the number equal to the fourth power plus or minus the cube plus the square. Knowing this, the problem is solved thus:

Add $2b$ to the coefficient of x^2. The square of half of this [$2bx^2$] is b^2x^4. Divide this by 64 and you will have $\frac{1}{64}b^2x^4$. Hence you see that, in order to have a root, you have added $\frac{1}{64}b^2$ [39] as the coefficient of x^4 and $2b$ as the coefficient of x^2. Therefore you add the same to $x^4 - 3x^3$ and you will have $(\frac{1}{64}b^2$ [40] $+ 1)x^4 - 3x^3 + 2bx^2$. For this to have a square root, the product of the extremes must be as much as the square of one-half the mean quantity. This is half of $1\frac{1}{2}x^3$ which, being squared, is $2\frac{1}{4}x^6$; and $(\frac{1}{64}b^2 + 1)x^4$ times $2bx^2$ produces

[39] 1570 and 1663 have $\frac{1}{16}x^2$.

[40] The text has $\frac{1}{46}$.

$(\frac{1}{32}b^3 + 2b)x^6$. For x^4 multiplied by x^2 produces x^6. You have, accordingly,

$$(\tfrac{1}{32}b^3 + 2b)x^6 = 2\tfrac{1}{4}x^6.$$

Therefore, as[41] x^6 is equal to x^6, so their coefficients are equal. Hence

$$\tfrac{1}{32}b^3 + 2b = 2\tfrac{1}{4}$$

wherefore

$$b^3 + 64b = 72,$$

and b equals $\sqrt[3]{\sqrt{11{,}005\frac{1}{27}} + 36} - \sqrt[3]{\sqrt{11{,}005\frac{1}{27}} - 36}$. Add twice this to both sides for [the coefficient of] x^2. The square root, then, of one side is $8 + x$ with a coefficient of the value of b and on the other x^2 with a coefficient equal to the square root of 1 plus $\frac{1}{64}$ the square of this value minus x with a coefficient of the square root of twice this value.

Problem XII

If someone says

$$x^4 + 3 = 12x,$$

add $2bx^2 + b^2$. It is clear that, without [3] the constant, this would have a square root. So we add $2bx^2$ to the other side and, as a constant, $b^2 - 3$. You will have these parts: $x^4 + 2bx^2 + b^2$, and $2bx^2 + 12x + b^2 - 3$. Therefore having multiplied the parts,[42] you have

$$2b^3 = 6b + 36$$

and

$$b^3 = 3b + 18,$$

wherefore b equals 3. Hence the parts are $x^4 + 6x^2 + 9$ and $6x^2 + 12x + 6$, and $x^2 + 3$, the square root of the first, will be equal to $x\sqrt{6} + \sqrt{6}$, and the value of x will be $\sqrt{\sqrt{6} - 1\frac{1}{2}} + \sqrt{1\frac{1}{2}}$.

Problem XIII

Find a number the fourth power of which plus twice its cube is 1[43] more than the number. Let

$$x^4 + 2x^3 = x + 1.$$

[41] 1570 and 1663 omit this word.

[42] I.e., having taken $(12x/2)^2$ and $2bx^2(b^2 - 3)$.

[43] 1663 omits this figure.

This leaves no room for a root to be subtracted or for division.[44] But in accordance with the first rule you say, Find three proportional numbers the ratio of the sum of which to the sum of the second and third is the same as that of the sum of the second and third to the first. Let these be 1, x, and x^2. You will then have

$$x^4 + 2x^3 + x^2 = x^2 + x + 1.$$

Hence, having removed the common square, we will have

$$x^4 + 2x^3 = x + 1.$$

We already know the ratios of the quantities, since the whole times the first equals the square of the sum of the second and third. Hence this sum must be divided according to a proportion having a mean and two extremes. The smaller part of this is 1, so the remainder (which is the greater part) is $\sqrt{1\frac{1}{4}} + \frac{1}{2}$ and this is equal, by supposition, to $x^2 + x$. Hence the quantities are (1st) 1; (2d) $x\left(\sqrt{\sqrt{1\frac{1}{4}} + \frac{3}{4}} - \frac{1}{2}\right)$; (3d) $x^2\left(\sqrt{1\frac{1}{4}} + 1 - \sqrt{\sqrt{1\frac{1}{4}} + \frac{3}{4}}\right)$.

Hence the fourth power of the middle term (which is x) plus $2x^3$ is equal to $x + 1$, and is $\sqrt{\sqrt{1\frac{1}{4}} + \frac{3}{4}} + \frac{1}{2}$. By this you know the methods for these rules if you have paid careful attention to the examples and the operations.

[44] *Hic non datur locus radici subtrahendae, nec divisioni.*

CHAPTER XL

On Forms of General Propositions Pertaining to the Great Art, and on Rules Which Are Out of the Ordinary, and on Solutions of a Nature Different from Those Which Have Been Spoken Of

1. If the cube is equal to the second power and constant and the coefficient of x^2 is subtracted from the solution, the remainder will be the solution for the cube and the same number of squares equal to a constant which bears the same ratio to the first constant as the second solution bears to the first.

For example,

$$x^3 = 2x^2 + 1\tfrac{17}{64}.$$

X is $2\frac{1}{4}$. I say that if [from this] you subtract 2, the coefficient of x^2, there will remain $\frac{1}{4}$, the solution for

$$y^3 + 2y^2 = \tfrac{9}{64},$$

which number is in the same ratio to $1\frac{17}{64}$, the constant of the first equation, as $\frac{1}{4}$, the second solution, is to $2\frac{1}{4}$, the first. Here is a demonstration of this:

DEMONSTRATION

Let AB be the first solution and let AC be the coefficient of x^2. BC will be the value for a certain cube and as many squares as the number AC equal to some constant, which is E. Let D be the number which, with as many AB's as the number AC, is equal to AB^3. Since, therefore, AB^3 equals the product of $(AC + CB)$ and

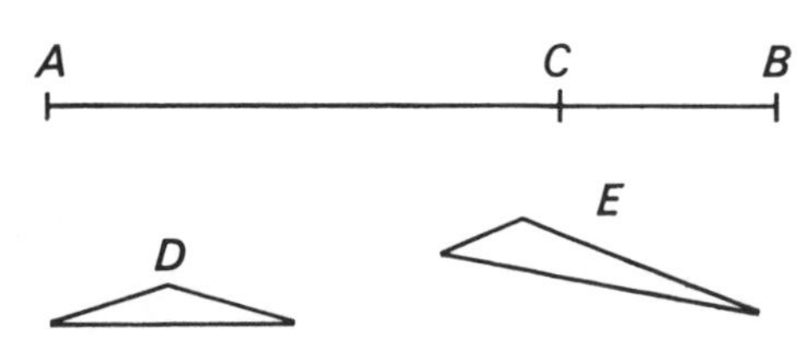

AB^2, and likewise the product $(AC \times AB^2)$ plus the number D, D is equal to the product $(CB \times AB^2)$. Likewise $BC^3 + (AC \times BC^2)$ is equal to the number E and, also, the product $(AB \times BC^2)$. Therefore $AB \times BC^2$ is equal to E. Then

$$(BC \times AB^2):(AB \times BC^2) = AB:BC,$$

according to the demonstrations in [our] seventh [book] on Euclid.[1] The ratio, therefore, of D to E is that of AB to BC, which was to be proved. It follows, likewise, by changing the proportions of the constants of the equations, that these are to their own proper solutions as the difference of their solutions to the coefficient of x^2.

2. If the cube and square are equal to a constant, and [another] cube is equal to a square with the same coefficient and the same constant, the ratio of the sum of the first solution and the coefficient of x^2 to the difference between the second solution and the coefficient of y^2 will be the square of the ratio of the second solution to the first. Thus, if I say

$$x^3 + 3x^2 = 20$$

and

$$y^3 = 3y^2 + 20,$$

x equals 2 in the first and y equals $\sqrt[3]{11 + \sqrt{120}} + \sqrt[3]{11 - \sqrt{120}} + 1$ in the second. I say that if you add 3, the coefficient of x^2, to 2, the first solution (which makes 5) and subtract this same 3 from the second solution (making $\sqrt[3]{11 + \sqrt{120}} + \sqrt[3]{11 - \sqrt{120}} - 2$), the ratio of 5 to this root is as the square of $\sqrt[3]{11 + \sqrt{120}} + \sqrt[3]{11 - \sqrt{120}} + 1$, the second solution, to the square of 2, the first solution. Here is the demonstration of this:

Demonstration[2]

Let BC be the first solution, AB the second, and AD the common coefficient of the second power. Since, then, AB^3 equals the product of $(AD + DB)$ and AB^2, and AD^3 is the coefficient of the second power, the product of

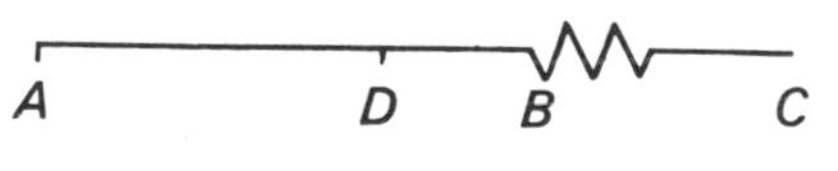

[1] 1545 has *ex demonstratis in septimo super Euclidem*; 1570 and 1663 have *per 143 libri de propor. colligitur, et in lib. Alizae.* The 143d proposition in the *De Proportionibus* reads thus: *Si linea in duas partes dividutur, corpora, quae fiunt ex una parte in alterius quadratum mutuo aequalia sunt corpori, quod sit ex tota linea in superficiem unius partis in alteram.* The reference to the Aliza book appears to be to its Chapter IX, entitled "Quomodo ex

DB and AB^2 will equal the constant of the equation and will, therefore, equal $BC^3 + (AD \times BC^2)$. Hence the product of BD and AB^2 is equal to $(AD + CB)$ times CB^2. Hence, according to XI, 34 and VI, 7 of the *Elements*,[4]

$$(AD + CB):BD = AB^2:BC^2,$$

a duplicate ratio or proportion.

3. If an equation of the square equal to the cube and number is converted to one of the first power equal to the cube and number, the second solution must be added to or subtracted from one-third the coefficient of x^2 in order to derive the first.[5] The method is this: Take the difference between the constant of the given equation and twice the cube of one-third the coefficient of x^2 and let this be the number which, with y^3, is equal to as many y's as one-third the square of the coefficient of x^2. Then to derive the first solution, the second having been found, add one-third the coefficient of x^2, if the number is greater than twice the cube of one-third the coefficient of x^2, or subtract it if it is less than twice the cube of one-third the coefficient of x^2, and the difference or the sum is the first solution.[6]

For example,

$$x^3 + 80 = 9x^2.$$

Twice the cube of 3, one-third the coefficient of x^2, is 54. The difference between this and 80 is 26. Hence

$$y^3 + 26 = 27y.$$

---

quacumque linea constituantur duo parallelipeda non maiora quarta parte cubi lineae propositae"; see also the discussion in Chapter XXXIX, entitled "De dividendis duabus lineis notis secundum proportionem mutuam reduplicatam iuxta partes datas."

[2] In the original, the diagram is lettered thus: A C D B

I have revised to the form shown above in order to avoid inviting the futile exercise of trying to reconcile the geometric results which the original diagram yields with the literal demonstration which the text spells out. The length shown in the revised form for BC is not, of course, intended to be compared with that shown for AB.

[3] The text has AB.

[4] 1570 and 1663 reverse these citations.

[5] 1570 and 1663 add: "This comes from those [matters] which pertain to Chapter VII."

[6] I.e., if $ax^2 = x^3 + N$, $y^3 + [N \sim 2(a/3)^3] = (a^2/3)y$, and $x = a/3 + y$ if $N > 2(a/3)^3$, or $x = a/3 - y$ if $N < 2(a/3)^3$.

But 27 is one-third the square of 9. Hence the second solution is 1 which, added to 3, one-third the coefficient of x^2, makes 4 the first solution, since the constant, which is 80, is greater than twice the cube of one-third the coefficient of x^2, which is 54.

Another example:

$$x^3 + 5 = 6x^2.$$

Square 6, making 36, one-third of which is 12, the coefficient of y. Then subtract 5, the constant of the equation, from 16, twice the cube of 2, one-third the coefficient of x^2. The remainder is 11. Hence

$$y^3 + 11 = 12y,$$

wherefore y equals 1. Subtract, therefore, 1 from 2, one-third the coefficient of x^2, because the number is less than twice the cube of one-third the coefficient of x^2. This leaves the solution for

$$x^3 + 5 = 6x^2.$$

Demonstration

Here is a demonstration of this: Let AB be the coefficient of x^2, which is 9, and let AC be the value of x. The cube of this plus 80 is equal to the product of AB and AF, the square of AC. Let AD, DE, and EB each be one-third of AB, and let AG be a rectangle[7] and [equal to] one-third AB^2, in accordance with VI, 1 [of the *Elements*.] Since, then

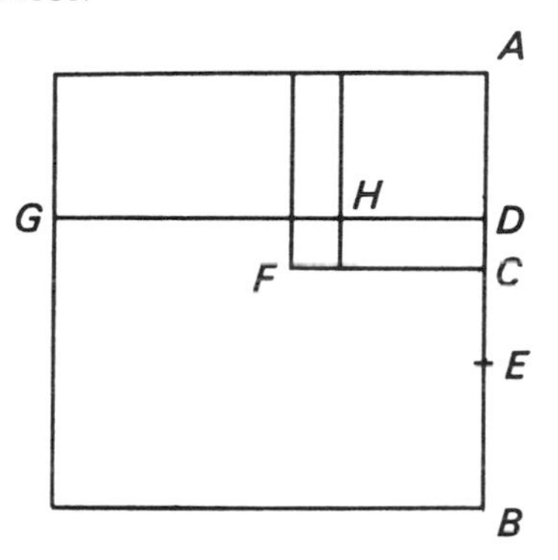

$$BA \times AF = AC^3 + 80$$ [8]

$$BC \times AF = 80.$$

Therefore

$$BD \times AF [=] 80 + (CD \times AF).$$

Having subtracted, therefore, $BD \times AH$,[9] which is $2AD^3$, [from $BD \times AF$], there is left BD times the gnomon [or] $26 + (CD \times AF)$.

[7] *superficies aequidistantium laterum.*

[8] The text has $BA \times AC = AF^3 + 80$.

[9] The diagram in the text has an L instead of an H; the L, moreover, is misplaced to appear along the top of the diagram.

And BD times the gnomon is equal to $4(CD \times AH) + 2(AD \times HF)$ because the lines BE, ED, DA, and DH and the other supplements are equal to each other. Therefore,

$$4(CD \times AH) + 2(AD \times HF) = 26 + (CD \times FA).$$

And

$$CD \times FA = CD^3 + 2(AD \times FH) + (CD \times AH).$$

Therefore, subtracting $(CD \times AH) + 2(AD \times HF)$ from both [sides of the second preceding equation],

$$3(CD \times AH) = CD^3 + 26,$$

and

$$3(CD \times AH) = CD \times AG$$

since DH equals one-third DG. Hence $CD \times AG$ (which is one-third AB^2) equals $CD^3 + 26$.

Even though the solution for a problem may come out with many terms, a solution can usually be hoped for, for this is nearly always a result of poor handling and it can generally be solved by reducing to fewer and known terms. But if [a problem] resolves itself into an equation with few but irregular terms,[10] its solution never comes to light in a general form since it always happens, in this type of equation, that there is no universal rule[11] for finding a solution. So, for instance, if it became an equation with the terms x^5, x^2, x, and N.

5. While this is true in all cases, it is particularly so in weighty geometrical problems. It is customary to gather a number of these together, so that you may anticipate others. You solve the less difficult ones with the help of this book and, in the end, reduce these solutions into rules of method and, proceeding step by step with their help through the precepts and rules relating to the unknown, you arrive at something beyond these [now] known rules, from which the solution will clearly appear.

6. Beyond these solutions, moreover, certain others appear, whose number is infinite. None of them is in general use, but those that are most frequent are of three kinds. [They may belong to] some particular rule, as is shown in the sixth book or, more [generally], in all the rules on quantities in continued proportion, as might well be expected.

[10] *ad capitulum paucarum sed inaequalium denominationum.*

[11] 1545 has *universalem . . . regulam*; 1570 and 1663 have *generalem . . . regulam.*

Others, however, derive from a repeated operation or a mixture of rules or types of equation as when, in order to solve a problem, you resort to several rules or types of equation. You had an example of this, forgetting about others, in the fourth problem of Chapter XXXV of this book when you followed out the problem to the end. It is also very clear in the second problem of Chapter XXXI of this book.[12] In the third class you will have various solutions when you employ the rules and equations using mixed quantities instead of integers as, if I say, Divide $\sqrt{8 - \sqrt{2}}$ into two parts, each of which multiplied by the root of the other gives certain numbers the sum of which is 4. This operation leads to an incongruous[13] quantity.

7. The nature of the product of [one of] the parts of a number and the root (square, cube, or any other sort) of the other part is similar to that of a cube or fourth power, except that the quantity must be assumed to be closer to the greater than to the less.[14] For example, if someone says, Divide 10 into two parts one of which times the square of the other is 9, and afterwards you wish to say, Divide some other number into two parts one of which multiplied by the square of the other is 18, then you see that such a product belongs to the class of the cube. [You might think,] therefore, since the proportion is the same, that this will come from 20, which is twice 10, just as 18 is twice 9.[15] But since it is of the nature of a cube, we look for two proportional terms between 10 and 20. These are $\sqrt[3]{2000}$ and $\sqrt[3]{4000}$. Therefore the number being sought is $\sqrt[3]{2000}$, one part [of which] is $\sqrt[3]{2}$ and the other $\sqrt[3]{1458}$.[16] Multiplying $\sqrt[3]{1458}$ by the square of $\sqrt[3]{2}$ gives $\sqrt[3]{5832}$, which is 18. I say, therefore, if you are asked to divide 10 into two parts, the product [of one] of which and the square root of the other is 12, that this has a cubic ratio. Hence if we were to say, You are to find a number the parts of which multiplied by their mutual

[12] 1545 has *Exemplum habes, praeter reliqua, in quarta quaestione capituli 35*[1] *huius libri, ubi eam quaestionem ad finem deduxeris, et expressius etiam in secunda quaestione 31, capituli huius;* 1570 and 1663 have *Exemplum habes superius, Capite 35, Quaestione 4. et Capite 31, Quaestione 2 huius, sed oportet perficere.*

[13] *absonam.*

[14] *Natura producti ex partibus numeri in* ℞ *quadratam vel cubam vel alterius generis partis reliquae, est de genere cubi, vel qdqdrati, excepto quod quantitas sumenda est proximior maxime, non minori.*

[15] The grammar of the original is somewhat confusing: *quia igitur, si proportio esset eadem, fieret hac ex 20, quod est duplum 10, ut 18 est duplum 9.*

[16] The text omits indication that this is a cube root; it uses only the simple radical sign.

roots produce 24 and from the first parts you are to find others between 10 and 20, which are in the same ratio as 9 and 18, you take two mean proportional terms in cubic ratio and the greater of these, which is $\sqrt[3]{4000}$, is the term sought for, for one part is $\sqrt[3]{4}$ and the other $\sqrt[3]{2916}$. Multiply, therefore, one by the square root of the other and there are produced $\sqrt[3]{5832}$ and $\sqrt[3]{216}$, which are 18 and 6, and the sum of these is 24.

8. Any equation of the cube equal to the first power and constant is convertible into a similar one in which the coefficient of y is formed by dividing the coefficient of x by the constant of the equation, and the [new] constant is the square root of that which comes from dividing 1[17] by the constant of the equation.[18] For example,

$$x^3 = 6x + 2.$$

Divide 6, the coefficient of x, by 2, the constant of the equation, and 3 is produced as the coefficient of y. Divide, again, 1 by 2, the constant of the equation, and $\frac{1}{2}$ results, the square root of which is the [new] constant. So, also, for the two remaining examples [shown in the margin].

$$x^3 = 6x + 2 \qquad y^3 = 3y + \sqrt{\tfrac{1}{2}}$$

$$x^3 = 4x + 4 \qquad y^3 = y + \tfrac{1}{2}$$

$$x^3 = 6x + 9 \qquad y^3 = \tfrac{2}{3}y + \tfrac{1}{3}$$

The discovery, however, of one solution from the other is very difficult. Nevertheless I say that, having the second solution, the first will be [obtained] by multiplying 1 plus the square root of the value of y[19] by[, on the one hand,] the cube and, on the other, the first power and constant, [all these terms being from the first equation]. Then you add to both [results] as many squares as come from dividing 1 by four times the square of the second solution. You will then have, on one side, fourth, third, and second powers with a square root which will be [in the form of] $bx^2 + ax$ and, on the other, a square, first power, and number, likewise having a root which will be [in the form of] $x + n$. From this you can derive the

x^3	$6x + 9$
$x + 1$	$x + 1$
$x^4 + x^3$	$6x^2 + 15x + 9$
$x^4 + x^3 + \frac{1}{4}x^2$	$6\frac{1}{4}x^2 + 15x + 9$
$x^2 + \frac{1}{2}x$	$2\frac{1}{2}x + 3$

[17] 1545 has R̸ *unitatis divisae per numerum aequationis*; 1570 and 1663 have R̸ *eius quod provenit divisa monade per numerum aequationis.*

[18] I.e., if $x^3 = ax + N$, $y^3 = (a/N)y + \sqrt{1/N}$.

[19] 1545 has *veruntamen dico, quod habita secunda aestimatione, ipsa erit* R̸ *numeri rerum multiplicandarum cum unitate*; 1570 and 1663 substitute *cum monade seu uno* for *cum unitate.*

solution through the rule. In the third case, for example, you have 1 as the solution for y. To derive the solution for x, multiply x^3 and $6x + 9$ by $1x + 1$, 1 being added in accordance with the rule and $1x$ being used because [1 is] the square of the value of y, which is also 1. You will [then] have

$$x^4 + x^3 = 6x^2 + 9 + 15x.$$

Now add $\frac{1}{4}x^2$ to both sides, [$\frac{1}{4}$] being what results from dividing 1 by four times the square of the value of y,[20] and you will have parts both of which have square roots. Therefore x equals 3.

WRITTEN IN FIVE YEARS, MAY IT LAST
AS MANY THOUSANDS[21]
THE END OF THE *GREAT ART* ON
THE RULES OF ALGEBRA
BY GIROLAMO CARDANO[22]

Printed in Nürnberg by Joh. Petreius in the year 1545[21]

[20] *et est quod provenit semper divisa unitate per quadruplum quadrati numeri positionum additarum.*

[21] 1570 and 1663 omit.

[22] 1663 omits.

APPENDIX

Portions of Euclid's Elements *Cited by Cardano*[1]

Book I

Axiom 1	Things equal to the same thing are also equal to one another.
Axiom 3	If equals be subtracted from equals the remainders are equal.
Proposition 35	Parallelograms which are on the same base and in the same parallels are equal to one another.
Proposition 42	To construct, in a given rectilineal angle, a parallelogram equal to a given triangle.
Proposition 43	In any parallelogram the complements of the parallelograms about the diameter are equal to one another.
Proposition 44	To a given straight line to apply, in a given rectilineal angle, a parallelogram equal to a given triangle.
Proposition 47	In right-angled triangles the square on the side subtending the right angle is equal to the squares on the sides containing the right angle.

Book II

Definitions	Any rectangular parallelogram is said to be contained by the two straight lines containing the right angle. And in any parallelogrammic area let any one whatever of the parallelograms about its diameter with the two complements be called a gnomon.
Proposition 3	If a straight line be cut at random, the rectangle contained by the whole and one of the segments is

[1] The texts here used are from T. L. Heath's *The Thirteen Books of Euclid's Elements* (2d ed., rev., Cambridge University Press, 1926) and are reproduced here by permission of the publisher.

equal to the rectangle contained by the segments and the square on the aforesaid segment.

Proposition 4 — If a straight line be cut at random, the square on the whole is equal to the squares on the segments and twice the rectangle contained by the segments.

Proposition 4, Corollary — From this it is manifest that in square areas the parallelograms about the diameter are squares.[2]

Proposition 5 — If a straight line be cut into equal and unequal segments, the rectangle contained by the unequal segments of the whole together with the square on the straight line between the points of section is equal to the square on the half.

Book V

Proposition 2 — If a first magnitude be the same multiple of a second that a third is of a fourth, and a fifth also be the same multiple of the second that a sixth is of the fourth, the sum of the first and fifth will also be the same multiple of the second that the sum of the third and sixth is of the fourth.

Proposition 11 — Ratios which are the same with the same ratio are also the same with one another.

Proposition 12 — If any number of magnitudes be proportional, as one of the antecedents is to one of the consequents, so will all the antecedents be to all the consequents.

Proposition 19 — If, as a whole is to a whole, so is a part subtracted to a part subtracted, the remainder will also be to the remainder as whole to whole.

Proposition 22 — If there be any number of magnitudes whatever, and others equal to them in multitude, which taken two and two together are in the same ratio, they will also be in the same ratio *ex aequali*.

Proposition 24 — If a first magnitude have to a second the same ratio as a third has to a fourth, and also a fifth have to the second the same ratio as a sixth to a fourth, the first and fifth added together will have to the second the same ratio as the third and sixth have to the fourth.

[2] Heath does not give this corollary but notes it (I, 381) as an undoubted interpolation which appears in the Greek text.

Book VI

Proposition 1	Triangles and parallelograms which are under the same height are to one another as their bases.
Proposition 7	If two triangles have one angle equal to one angle, the sides about other angles proportional, and the remaining angles either both less or both not less than a right angle, the triangles will be equiangular and will have those angles equal, the sides about which are proportional.
Proposition 16	If four straight lines be proportional, the rectangle contained by the extremes is equal to the rectangle contained by the means; and, if the rectangle contained by the extremes be equal to the rectangle contained by the means, the four straight lines will be proportional.
Proposition 17	If three straight lines be proportional, the rectangle contained by the extremes is equal to the square on the mean; and, if the rectangle contained by the extremes be equal to the square on the mean, the three straight lines will be proportional.

Book VII

Proposition 17	If a number by multiplying two numbers make certain numbers, the numbers so produced will have the same ratio as the numbers multiplied.

Book XI

Proposition 31	Parallelepipedal solids which are on equal bases and of the same height are equal to one another.
Proposition 32	Parallelepipedal solids which are of the same height are to one another as their bases.
Proposition 34	In equal parallelepipedal solids the bases are reciprocally proportional to the heights; and those parallelepipedal solids in which the bases are reciprocally proportional to the heights are equal.

INDEX

A CATALOG OF SELECTED

DOVER BOOKS

IN SCIENCE AND MATHEMATICS

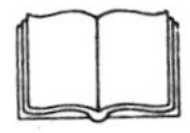

A CATALOG OF SELECTED
DOVER BOOKS
IN SCIENCE AND MATHEMATICS

NUMERICAL METHODS FOR SCIENTISTS AND ENGINEERS, Richard Hamming. Classic text stresses frequency approach in coverage of algorithms, polynomial approximation, Fourier approximation, exponential approximation, other topics. Revised and enlarged 2nd edition. 721pp. 5⅜ × 8½.
65241-6 Pa. $14.95

THEORETICAL SOLID STATE PHYSICS, Vol. I: Perfect Lattices in Equilibrium; Vol. II: Non-Equilibrium and Disorder, William Jones and Norman H. March. Monumental reference work covers fundamental theory of equilibrium properties of perfect crystalline solids, non-equilibrium properties, defects and disordered systems. Appendices. Problems. Preface. Diagrams. Index. Bibliography. Total of 1,301pp. 5⅜ × 8½. Two volumes. Vol. I 65015-4 Pa. $12.95
Vol. II 65016-2 Pa. $12.95

OPTIMIZATION THEORY WITH APPLICATIONS, Donald A. Pierre. Broad-spectrum approach to important topic. Classical theory of minima and maxima, calculus of variations, simplex technique and linear programming, more. Many problems, examples. 640pp. 5⅜ × 8½. 65205-X Pa. $13.95

THE MODERN THEORY OF SOLIDS, Frederick Seitz. First inexpensive edition of classic work on theory of ionic crystals, free-electron theory of metals and semiconductors, molecular binding, much more. 736pp. 5⅜ × 8½.
65482-6 Pa. $15.95

ESSAYS ON THE THEORY OF NUMBERS, Richard Dedekind. Two classic essays by great German mathematician: on the theory of irrational numbers; and on transfinite numbers and properties of natural numbers. 115pp. 5⅜ × 8½.
21010-3 Pa. $4.95

THE FUNCTIONS OF MATHEMATICAL PHYSICS, Harry Hochstadt. Comprehensive treatment of orthogonal polynomials, hypergeometric functions, Hill's equation, much more. Bibliography. Index. 322pp. 5⅜ × 8½. 65214-9 Pa. $9.95

NUMBER THEORY AND ITS HISTORY, Oystein Ore. Unusually clear, accessible introduction covers counting, properties of numbers, prime numbers, much more. Bibliography. 380pp. 5⅜ × 8½. 65620-9 Pa. $8.95

THE VARIATIONAL PRINCIPLES OF MECHANICS, Cornelius Lanczos. Graduate level coverage of calculus of variations, equations of motion, relativistic mechanics, more. First inexpensive paperbound edition of classic treatise. Index. Bibliography. 418pp. 5⅜ × 8½. 65067-7 Pa. $10.95

MATHEMATICAL TABLES AND FORMULAS, Robert D. Carmichael and Edwin R. Smith. Logarithms, sines, tangents, trig functions, powers, roots, reciprocals, exponential and hyperbolic functions, formulas and theorems. 269pp. 5⅜ × 8½. 60111-0 Pa. $5.95

THEORETICAL PHYSICS, Georg Joos, with Ira M. Freeman. Classic overview covers essential math, mechanics, electromagnetic theory, thermodynamics, quantum mechanics, nuclear physics, other topics. First paperback edition. xxiii + 885pp. 5⅜ × 8½. 65227-0 Pa. $18.95